Living
Wild

Living Wild

The Secrets of Animal Survival

by David Robinson

National Wildlife Federation

Contents

QL751
.R68
1980

Introduction

Before you begin this book, you may enjoy turning back to the Contents page for a moment to test your knowledge of animal behavior, by playing a word-association game with the clues you find there.

Take the title of the first chapter, The Breadwinners. What animals do you associate with that description? The hardworking bees and ants? A cheetah stalking a gazelle?

How about The Defenders? Who comes to mind? The well-armed skunk?

And which animals qualify as The Travelers? Canada geese? Monarch butterflies?

All of these answers are, of course, correct. And they are probably the responses many people would make. Yet they do not give a complete picture, for every animal plays a variety of roles, some as many as eight or nine, to meet the daily challenges of survival.

In planning this book, our editors decided to focus on these varied roles as a useful way of sharing with you some of the fascinating facts about animal behavior gleaned from the work of researchers and wildlife managers throughout the world.

You will find that looking at animal behavior in this admittedly anthropomorphic fashion provides a checklist for reviewing your knowledge of each animal. For example, it reminds us that Canada geese are not only prodigious travelers who can fly from Mexico to Canada; they are also breadwinners capable of identifying grainfields adjacent to water where they may touch down to feed en route.

And these are only two of their survival roles. As defenders, they post sentinels while the rest of the flock rests from their arduous journey. As builders, they construct a ground nest on the cold tundra and the female warms it with a lining of down plucked from her breast. As courtship strategists, the gander vigorously pursues the goose until they mate for life; and as parents, he stands guard while she incubates the eggs and together they rear their goslings.

In all these exquisitely timed activities, the Canadas move as a flock, a feat possible only for skillful cooperators and communicators. And if they find that last year's inviting marshy rest stop has been paved over to become this year's shopping mall, they prove themselves to be hardy, successful innovators. So if you play the word game on the Contents page, the role associated with the Canada goose, as with many another species, would have to be "all of the above."

Understanding what animals are doing—and are likely to do next—adds immeasurably to the enjoyment that birdwatchers, backpackers, photographers, hunters, and fishermen find in their outings. Sometimes it adds to their safety. It can also be useful to the suburban family whose attic has been invaded by raccoons.

But knowledge of the full range of each animal's specific behavior patterns is an absolute necessity for refuge managers, lumber company wildlife biologists, zoo keepers, and a roster of other wildlife professionals. In the early years of my own career in Arizona and Colorado, I learned that making wise decisions for herds of elk or flocks of migrating ducks keeps all wildlife managers dependent on an exchange of information with researchers in the laboratory and in the field. And now, after two decades of working with the National Wildlife Federation and with others who are trying to steer the nation in the formation of sound conservation policies, I am even more aware of our need for current knowledge of how animals are adapting to the pressures of civilization.

So as our author, David Robinson, transports you into the dark burrows, the gurgling sea caves, and the windswept aeries of the planet to see how animals actually live, I hope you will sense his indirect tribute to all the researchers whose work has made this book possible. On skis, in helicopters or rowboats, these determined fact-finders clad sometimes in fur parkas, sometimes in wetsuits, patiently pursue their quarry. Watching, waiting, recording the animals' every move, the researchers' days—and nights—veer from hours of boredom to moments of wild excitement. Eventually, they may uncover or pin down one more previously elusive fact. The breakthrough is often recorded in stunning photographs, some of which fill the pages of this book.

Leafing through it, I was pleased to see how solidly the photographs support the role concept of animal behavior, and how they document in fine detail the rich array of skills that animals have developed to fulfill the roles. In The Breadwinners, for example, we meet both the gentle vegetarian grazers and browsers and the stealthy, silent predators who prey upon them and upon each other: the bird of prey that strikes from the sky, the stalkers, the ambushers, the trappers, the poisoners. Working equally hard to secure their next meal are the tool users, the chasers, the divers, and last—of necessity—nature's clean-up crew, the scavengers. In each chapter, with every turn of the page we witness a different animal technique or stratagem.

I urge readers to yield as I did to the temptation to "read" the pictures first. Taking our cues from the caption headlines to identify the specific skills depicted, we discover the amazing versatility which wild things have developed in their drive to survive. With Bob Hynes' artwork taking us where the camera cannot readily go, this pictorial introduction to the book can only heighten our respect for the cast of characters which the author brings on stage in his text.

And what a cast it is. Like a drama critic's review of a virtuoso performance by the entire animal kingdom, David Robinson's insightful prose rings with applause and shouts of bravo. We could have found no more appreciative interpreter of the researchers' findings nor of their versatile subjects.

I am particularly glad that Dave has given us a book that can be enjoyed by all members of the family. I can imagine younger readers saluting many an old friend whom they have met in our children's magazines and delighting in learning more about them. How a mother elephant tells her calf it's time for a mudbath. Exactly how a crow steals the bait from an ice fisherman's submerged line. How a flying fish flies without wings.

At the same time, older members of the family will appreciate the way these vignettes and hundreds like them are used to establish the known frontiers of all animal behavior. Whether the author is locating an event in the solar system or pegging it in geologic time or along the evolutionary trail, he helps us to see the smallest things in the largest possible framework. His own exuberant air of discovery nourishes our sense of wonder.

I am convinced that every one responds with some degree of curiosity and pleasure to the mystery and actions of the wild things around us—which is to say that we are all animal behaviorists at heart. Indeed, I think we may safely call the study of animal behavior man's oldest intellectual pursuit. Despite the sophisticated gear used by today's wildlife research teams—from underwater vehicles to tranquilizers and radio-tracking devices—today's observers are almost certainly asking the same questions that were pondered by primitive men tens of thousands of years ago. What can this animal do? Why does it act this way? Where does it get its wisdom?

For our cave-dwelling hunter ancestors, finding answers to those questions was often a hair-raising, fateful matter on which they had to concentrate all their powers. Their objective was simply to live from one hunt to the next—to keep themselves and, in turn, the human race alive. Today's wildlife managers seek the information so they can help wild animals continue to live on this increasingly crowded planet.

The National Wildlife Federation is committed to keeping the beauty and excitement of wildlife in our lives and strives to enlist each succeeding generation in accomplishing that goal. This book will help if it arouses new interest in and respect for the marvels of animal behavior. As detail after detail of the secrets of animal survival continue to be painstakingly discovered, eventually the blurred picture of the whole panorama of nature—and our role in it—will come into focus.

Thomas L. Kimball
Executive Vice President
National Wildlife Federation

Eating Without Being Eaten

Hunger casts every wild
creature in the roles of
breadwinner and defender.

The Breadwinners

The little birds are making big pests of themselves, and the sleepy-eyed owl has had enough. Silent wings reach out and gather in the afternoon air, lifting the gray screech owl from its perch in a northern forest. Seeking a better place to get a day's rest, the night hunter moves out on slow, steady wingbeats through the trees, across a sunny meadow, out over the mirror of a placid pond. And there over the pond its life explodes in a burst of feathers; in an instant a dead owl is dropping toward its own watery reflection. What had been only a dot in the sky above the owl has suddenly become the feathered thunderbolt of a peregrine falcon, probably the fastest of all earth's creatures in its spectacular "stoop" or dive after prey. The owl never sensed the falcon's approach, never felt the talon that pierced its skull—and never splashed into its own reflection, for the falcon wheeled and seized it in midair.

Beneath another pond mirror far to the south, an alligator snapping turtle bulges from the oozy bottom like a lifeless rock. Its hook-beaked jaws stand agape and motionless, and inside the fearsome maw a wormlike jut of flesh on the turtle's tongue begins to wiggle. The minutes crawl by. Then a small, silvery fish swims past, spots the "worm," edges in for a closer look. SNAP! When the swirl of silt settles, there lies the same "rock," there squirms the "worm," and sooner or later, into the open jaws comes another feeder to be fed upon.

The peregrine fetched a meal at dazzling speed—at least 90 miles an hour. The snapper also dined, yet the 200-pound reptile moved scarcely a muscle and traveled not an inch. Between the two strategies, nature arrays a bewildering repertoire of answers to a command nearly all her creatures must obey: eat until you are eaten.

And so the owl becomes part of the peregrine that now feeds upon it. So does the black rat snake that had earlier become part of the owl; so does the mouse absorbed by the snake; so does the grass seed that was the mouse's final meal. We call this progression a food chain.

Food is the income of life from which a creature's expenditures of energy are paid. As the peregrine stoops to conquer, it whips its wing-tips on the way down, thus buying speed and paying for it with energy. The feather it lost in the kill must be replaced, and the new one will also cost energy. Digestion costs energy; elimination of unused material costs energy; it even costs energy to store energy in the form of fat. From the energy stored in the flesh of the owl, the falcon will more than replace what it spent.

Virtually all of the energy being passed from creature to creature comes from a single dynamo: the sun. The intricate process by which it produces flesh and bone to keep the hungry fed is worth reviewing before we explore more of the fascinating ways animals acquire their daily bread.

Every day the sun drenches this life-filled planet with more energy than earth can possibly use. Some glances off our canopy of air and is lost in space. Some heats the land, piles up the clouds, and stirs the air masses into a chowder we call weather. And some falls into the outstretched green hands of life's miracle-makers, the plants. With it the plants can manufacture their own food. The word for this magical process is photosynthesis.

It begins as a raindrop splats on a rock and seeps into the soil. There the moisture gathers up carbon, oxygen, hydrogen, nitrogen, and many more of life's basic building blocks. A rootlet drinks in the freighted drop and the plant's plumbing hoists it to a leaf which is also receiving carbon dioxide from the air.

The rain stops. The sun blazes down again. Now the plant's green molecules of chlorophyll work their alchemy at full speed. In a complex process not yet fully understood, they catch the sun's energy and use it to make food. From the carbon dioxide and the mineral-laden water the plant concocts it in the form of carbohydrates to be released as needed for the business of each cell. Thus plants become batteries charged with power from the sun, and almost everything that lives runs on that power.

Photosynthesis may seem to be a wasteful miracle. Only about one percent of the solar energy falling on a typical plant actually ends up being converted into plant matter. But on that one percent the whole animal kingdom depends—including the entire human race.

A bird of prey scans the Snake River Canyon for a meal.

Spotting a migrating teal duck a thousand feet below, a peregrine falcon can nose into a power dive, strike the duck in midair, and follow it to the ground; there a sharp twist of the falcon's notched beak breaks the victim's neck, completing the kill. At times the falcon hunts with her smaller mate, the tiercel, one flushing, the other attacking.

Despite its status as a top predator in its ecosystem, the species has become endangered by eating pesticide residues which accumulate as they are passed along in the flesh of all the creatures in its food chain.

Previous page: Lion, Thomson's Gazelle, Serengeti waterhole.

Grazers convert plants into flesh, bone, and beauty.

Urged on by hunger, bison (opposite), pronghorn (left), and a pika (above) select the plants they need to live another day. The bison's habit of yanking up sedges and grasses by the roots damages the range; but the pronghorn actually improves the pasture for cattle and sheep by its dainty nipping of plants and its preference for sagebrush, thistles, and noxious weeds. The pika, "the little haymaker," harvests 30 kinds of plants and dries them in the sun—ever alert for the swoop of a hawk.

Winter brings hard times for the big animals. The bison's enormous head snowplows through up to four feet of snow to graze; the pronghorn can paw through only inches of it. Now predators take the stragglers, while the pika—snug in its 50-pound cache of hay—eats and sleeps the winter away.

Breadwinning for many animals is a simple matter of grazing or browsing. Not for them the life-and-death drama of the predator's search for every meal; yet taking the sun power stored in a plant and converting it into animal protein is no mean feat. Unlike most animal cells, a typical plant cell wears a tough jacket of cellulose. Even a grazing animal's digestive system can't handle that without help. Thus the bison's jaw is studded with broad, heavy teeth that crush the tiny fortresses to a pulp. Down slides the pulp to the animal's rumen, the first of four chambers in its stomach. In the warm shelter of the rumen dwells a hidden herd of bacteria that completes the breakdown of the cellulose. Then as the bacteria are themselves digested, the bison also unwittingly culls from these microscopic organisms the protein that it needs.

The bison helps the bacteria, the bacteria help the bison. This kind of scratch-my-back-and-I'll-scratch-yours is called mutualism or symbiosis and many vegetarians depend on such arrangements for survival. Like the lordly bison, lowly termites carry a gutful of protozoans to digest their woody diet for them. Oddly, when a growing termite sheds its skin it also sheds its protozoans; it must reinfect itself by eating the feces of its broodmates. This may strike us fastidious humans as acceptable behavior for insects, yet our children may be shocked to see their pet rabbit's baby nibble its mother's droppings. The youngsters need to understand that rabbits too are born without the vital colony of protozoans and must obtain them in this fashion. The bunny is merely stocking up on a living survival kit.

Chomp, nip, sip. Browsers enjoy plants' varied menu.

A honeybee (opposite) sips from the nectar cup of a cosmos whose red flare signals its need of a pollinator. With honeybag full and back-leg baskets packed with the flower's protein-rich pollen, the bee ferries its cargo to the hive. Trying to feed the thousands of hungry mouths there, the bee will become exhausted and die within three weeks.

No such altruism attends the gorging of the four-inch caterpillar of the Cynthia silk moth (below left). Chewing its way toward a new life as one of the largest moths, its immediate goal is to split its skin by eating ailanthus leaves.

Nature's tallest browser (bottom right) sniffs its favorite tree, the acacia, before it tastes.

The giraffe's daily menu — some 30 pounds of twigs, leaves, thorns, pods, fruits, and galls — is swallowed whole and chewed later.

The bonanza of an ear of corn enables an eastern cottontail (below) to bear with better grace the long wait for winter's brown forage to turn to spring's green banquet. Like all herbivores, its fortunes rise and fall with those of the plants.

15

The vegetarians serve as a vital bridge from plants to man and other meat-eaters, nourishing us with nutrients from plants we cannot eat.

The grazers and browsers seem a prosaic lot. After all, their prey is a plant, a victim that can only stand rooted to the spot while animals amble up and bite it off. Yet even here the thrust and parry of combat is played out in a low-keyed way.

Some plants have evolved rather ingenious defenses: thorns, tough armor, a terrible taste, an evil smell. In response the plant eaters have devised some remarkable ploys. The acacia tree protects its leaves with an arsenal of needle-sharp thorns. But the narrow muzzles of the gerenuk and other African antelopes slip easily between them. Nut trees armor their seeds in shells that often call for our stoutest

This tense tableau will explode into pandemonium if one of the Thomson's gazelles breaks under the strain, giving the cheetah's brain the fleeing target it seems to need to trigger an attack. The elegant cat is accustomed to failing in four out of five attempts to raid a herd, for achieving total surprise in the grasslands is not easy — even for one of the world's best stalkers and fastest sprinters. Hunger will goad the cheetah to try again soon, perhaps first climbing a small hillock or leaping up onto a low tree limb to plan new strategy. Subtle variations in the cheetah's hunting ritual keep the gazelles wary. Next time the cat may prolong the crouching phase of its approach, the grass may be a bit higher, the herd less vigilant.

nutcrackers. So the nut-loving gray squirrel has developed chisel-like gnawing teeth backed up with a pair of jaws that can clamp down on a black walnut with an incredible 11 tons of biting pressure per square inch!

As the vegetarians eat their fill, their role too seems an inefficient one. Only about 10 or 15 percent of the sun's energy passed to them by the plants they eat is actually converted into body mass. Much of that energy is lost as body heat, dissipated into the air at the command of internal thermostats. So when the meat-eaters have killed and eaten their fill of the grazers and browsers, even less of the energy finds its way into the fabric of the killers. Again, much has vanished as heat from their bodies. In sunlight's erratic odyssey from the surface of the leaf, only about one thousandth of one percent of its energy finally makes its way into the living cells of the predator.

But what magnificent creatures those cells do build! Consider the cheetah as it stalks a gazelle herd across the African savanna. Its lean form hugs the ground as its sharp eyes scan the herd, searching for the sign of weakness or lameness that will mark one animal as an easy catch. Nothing on earth can outrun the cheetah as it sprints at 60 miles an hour, but it can only keep up that dazzling dash for less than a minute. To shrink the distance it must first hunker down in the grass, take one careful step forward, then another, creeping closer, closer.

There it goes! Its hind legs kick the blurred earth like pistons as the gazelles scatter in panic. The cheetah's feet fly, its muscles labor rhythmically, its haunches rise and fall with

Snaring prey with silk is sure-fire strategy for spiders.

One golden orb weaver (below) completes a winding sheet for a dying grasshopper, while another (right) hustles to do the same for a dragonfly that has just blundered into its web. Neither the grasshopper's defensive spitting of brown "tobacco juice" nor the dragonfly's own skill as a snarer of

every ground-gulping stride. Yet its head follows a relatively even line, for the eyes are constantly transmitting information about the fleeing quarry and must hold a steady bead on their target. This time the cheetah has crept close enough; its dash closes the remaining distance and wins it an ample meal. Next time it may spook the herd too soon—or even lose its footing at the last instant and tumble head over tail in one of the most spectacular pratfalls in nature.

That doesn't happen often, for the cheetah's paws are well fitted for running. Its claws are blunt and only partly retractable, for their main job is to hold the ground as a human sprinter's spikes do. When the cheetah closes in, its claws can grasp the luckless quarry long enough for its teeth to make the kill.

The cheetah's claws are like the wolf's. That isn't surprising, since cheetah and wolf both hunt by running, and both use their blunt claws to hold down a bone while sharp teeth and powerful jaws wrench off meat and gristle. But unlike the cheetah, the wolf rarely wins its race by a short dash but by dogging its quarry's tracks at a patient trot, a pace it can keep up for hours.

"Predator" is a vivid word. It conjures up epic struggles between mighty combatants. Yet the largest predator that ever lived is the 135-ton blue whale, and its prey is no bigger than the inch-long krill and even smaller plankton drifting ceaselessly with the seas. And its "epic struggle" is nothing more than a swim through this drifting swarm of minute plants and animals with its cavelike mouth open, then a push with its tongue to force the water out and seine the plankton from it with sieves of whalebone called baleen. Clearly there are many ways to be a predator, and not all of them involve savage combat. In fact, most breadwinners prefer to get a meal with as little risk as possible, so they favor animals that are smaller than themselves.

A few predators attack prey that is much larger than they are, but most of them team up to do so. The black-and-white killer whale is dwarfed by its blue whale cousin, yet killer whales have been seen mobbing the gentle leviathan, ripping great chunks out of its body

and finally leaving it in a cloud of bloody water, wounded perhaps unto death.

In the waters of the South Atlantic off Patagonia, the killer whale and the smaller sea lion play out the grim game of survival, the one as hunter, the other as prey. Sometimes the unseen drama beneath the placid waves has exploded before the startled stares of explorers and scientists. For the killer whale has developed the seemingly cruel habit of batting the 150-pound sea lion 30 feet into the air with its mighty tail.

Cruel? Surely the whale knows no such human emotion. It lofts the helpless sea lion for a purpose we can only guess—perhaps to soften it up for easier feeding, possibly to stun it for an easier kill. Only humans make moral judgments when they witness such scenes; only we can be humane. The wild predator is simply securing a meal.

The stage for the drama of breadwinning is not always of such heroic proportions, however. What we perceive as cunning, pathos, savagery, humor—all are often enacted on the stage of a single fallen leaf. At first glance, there is nothing epic about a small, fuzzy, short-legged spider sauntering across a forest floor. But watch it awhile, and it will reward you with drama aplenty. The leap of the jumping spider is so trigger-fast, neither you nor an unwary fly could see it coming. Beneath its eight eyes—the best in spiderdom, though they can identify prey no more than a foot away—hangs a pair of jaws tipped with hollow fangs. The tussle with the fly is brief as venom from glands in the spider's head shoots through the fangs and paralyzes the victim.

All spiders are on liquid diets; why then do they hunt the solid fly? Because it won't be solid for long. Enzymes from the spider's maxillary glands liquefy the fly's innards, enabling the tiny predator to suck them into its own stomach. Thus, spider digestion starts not in the spider but in the meal.

Another dangerous landing field for the fly is the lily pad. The slow-moving, dull-witted frog sunning itself thereon would offer no threat to the agile fly were it not for a short circuit which nature has devised between the frog's eyes and its brain. When the eyes see

insects in its hairy-legged "basket" could save them from the
spider's venomous bite. Extrud-
ing fresh, sticky silk from its
body, each spider turns its
paralyzed victim around and
around. With each prize safely
enshrouded, its captor will drag
the body to the hub of the web.
There it will dine as soon as
the insect's innards have been
liquefied by the venom's
enzymatic action.

the fly, the signal to strike bypasses the brain and triggers the tongue. Out it snaps, a sticky, stretchy retriever that can tack on a fly at two inches. Almost anything that moves and is not too big can trigger the frog's tongue—a grasshopper, a small snake, even another frog.

As we trace a food chain from one animal to another, we find ourselves climbing a pyramid; the sun-given energy dwindles and so do the numbers of animals involved. Thus, there are many more flies than frogs. And there are many more frogs than frog-eating herons. The survival of all creatures demands that this be so, for unless the prey far outnumber the predators, the food supply will be quickly eaten to extinction and the predators will soon follow by starvation.

As on the land, so in the sea: the energy pyramids rest on a foundation of creatures that number in the uncountable trillions of trillions. These are the plankton, the microscopic drifters that can cloud vast volumes of sea with their numbers. Among them are the diatoms, the "grasses of the sea" that catch the sunlight near the surface and turn the sun's energy into plant food as do the grasses of the land.

But what of the bottom dwellers? How is the sun's energy, harnessed near the surface, passed down into their murky domain? On the ocean's floor there is a constant rain of detritus from above—dead plankton, bits of flotsam, scraps unclaimed from a predator's kill. Thus brittle starfish and marine worms lying fathoms deep in their bed of ooze inherit in many forms their share of energy from a sun they never see.

In shallow waters a different miracle occurs as the oyster and other bivalves reap the bounty that sweeps over their beds from the land and its rivers. They process this bounty with one of the most amazing structures in all of nature. It is called the crystalline style.

When the oyster feeds, it filters a gumbo of plankton, detritus, sand, and any other minute particles that may be at hand. The oyster's gills and sensitive mouthparts sort out the bits of food and generate strings of mucus that pick up these bits like sticky conveyor belts. Then, slowly, these food-laden strings are drawn into the mouth and on into another chamber by a rotating gelatinous rod that reels them in like

Paralyzing prey with poison ensures leisurely meals.

Immobilized by the stinging tentacles of a sea anemone (top left), a helpless blenny awaits its fate. The petal-like tentacles soon close tightly over the little fish (top right), forcing it into the muscular mouth which slowly swallows it. In about 20 minutes, the anemone will open again and spit out any undigested parts.

A pinfish (opposite) meets a similar fate as it brushes against a Portuguese man-of-war, the free-floating cousin of the rock-bound anemone. One set of the jellyfish's yard-

long tentacles produces fine threads to hold the victim as it receives a paralyzing sting; meanwhile another set of tentacles spreads over the fish to digest it.

Not even the armored crab can escape the poison power of the octopus (above). How the poison gets inside the crab's shell isn't known, but once the octopus' sucker-clad arms enwrap the victim, it is helpless to resist the lethal dose. Paralysis, tremor, and death quickly follow.

The octopus also eats crabs and lobsters alive without bothering to poison them.

ropes on a windlass. This curious rod is the crystalline style, possibly nature's only rotating internal part.

As the rod slowly revolves, it seems to be working its way gradually out of its chamber and into the stomach proper. But as the head end of the rod enters the stomach, it dissolves into digestive juices while the other end of the rod continues to form. And when feeding stops, the whole rod can dissolve—a disposable miracle that the humble shellfish repeats whenever it eats.

Thus do some of the leftovers from life on land re-enter the energy pyramid as a succulent oyster, a delicacy to man and to many an ocean prowler as well.

On the ocean floor, there is no dearth of prowlers. The clam's stick-in-the-sand lifestyle shields it from some of them, and so does its rock-hard shell. But find a clam shell on a beach, and you'll probably notice a hole bored in it by a prowler who made its occupant a meal.

One of the most efficient of the ocean prowlers is the innocent-looking moon snail. When it finds a limpet glued to a rock, it creeps onto its shell and proceeds to rasp a hole in it. Then the soft parts of the limpet's flesh are drawn out the hole by the snail's powerful sucking mouth. The limpet's only defense is that they outnumber the moon snails.

The moon snail goes about its safecracking with quiet patience. But as we have already seen, the sea knows sudden violence as well. Divers who have watched from the safety of their stout, protective cages while two or three sharks rip into a swordfish or a small whale for dinner report it to be an awesome sight. But when conditions combine to attract and excite a mob of sharks, the divers witness a primitive spectacle. Spurred by the thrashing of a victim and the smell of blood in the water, these superb killing machines begin to dart and snap in mounting excitement. Their thrashing tails whip the sea to a boil. If the frenzy keeps up, they may even start shredding each other.

The feeding frenzy occurs in other life forms, but seldom does it escalate into cannibalism. Seeing an animal eat its own kind is repugnant to human sensibilities. Yet cannibalism is more common in nature than we realize.

Tables are sometimes turned in grim ambush scenario.

Scene I (left). A leopard seal exhibits amazing patience as it waits for hunger to drive Adelie penguins to feed in frigid Antarctic waters. Splash! When birds dive in, the seal dives too.

Scene II (below). Snap! Seal jaws close around a penguin head. With consummate skill the seal shakes the bird's body so furiously that the skin is peeled off before the flesh is devoured.

Scene III (above). A penguin-fattened Weddell seal clings in terror to another floe, its own ambush skills now useless as a pair of killer whales close in. If the seal doesn't panic and dive in, to almost certain death, the circling pair of Orcas may create waves that will sweep the seal off the ice. Failing that, they may bombard the floe from beneath until it breaks or until they have tilted the doomed seal into the sea.

Unless fiercely protected by their mother, some bear cubs may be killed and eaten by their own father. Behaviorists wonder whether this is a mechanism for controlling overpopulation when food is scarce rather than a feeding behavior. Whatever the reasons for it, cannibalism in any species is a chilling reminder of dark chapters in our own human history. And the reminder gains new immediacy as we learn that cannibalism is apparently an occasional response to overcrowding or other environmental pressures among the most manlike of nonhumans, the chimpanzees.

A few creatures such as the codling moth larvae habitually eat each other on contact. Thus even cannibalism can have its brighter side. If you find a codling moth larva—a worm —in an apple, you can relax. There almost certainly won't be another one.

Whether a species ever engages in cannibalism or not, it must have some sure way of protecting its turf. If two species find themselves competing for a single resource, one of them always has to go. They may fight to the death, one species may starve, or one may migrate to greener pastures, but it is one of the broad axioms of nature that two species cannot long occupy the same niche.

There is more to a niche than just a physical location or a food source. An animal's niche is the sum total of all the factors it needs to survive and reproduce its kind: the food it needs, the amount of water it must have, the climate it prefers, the shelter it requires, the presence or absence of others of its species and many other factors.

Yet it is possible for seven species of storks to feed within sight of each other in the wetlands of East Africa because the seven stick to separate menus. One grabs big fish, another catches little ones. A third eats mussels and snails, a fourth snatches insects. The fifth is a scavenger and the other two eat almost anything. The seven share the same habitat, but each extracts something different from it. Thus their niches are different and they can co-exist in peace.

Some creatures' niches are inflexible, firmly fixed down through the eons. Although such rigidity has worked well for the tortoise, the opossum, and other living ancients, it has been fatal to untold numbers of species. Most species survive because they are able to adapt to changing conditions and new opportunities; their niches have some "give" to them. There were ticks long before there were dogs, yet the tick accepted this newcomer as part of its niche and has thrived on it ever since. There were foxes before there were henhouses. But the fox has learned to exploit this rich new food resource.

Differences in eating patterns are dictated not only by an animal's external surroundings but by its internal metabolism, the rate at which each builds new cells and breaks down old ones to release energy and heat. A shrew starves to death in about three hours, a boa can last a year or more without a meal. Watch the frenetic shrew bustle about and the lethargic boa bask by the hour; in that difference of metabolism lies also one of the keys to the ascendancy of mammals.

To fuel this faster life-style, we mammals need more food than reptiles do. And we must get its nutrients into our systems as quickly as possible. That's why we chew a while before we swallow. Yet because we need constant oxygen for the chemical reactions that free our stored energy, we must be able to chew and breathe at the same time. Most reptiles cannot do this; their nostrils lead directly into the mouth. But a mammal's nasal passages are separated from its mouth by a secondary palate; it can chew and go right on breathing.

Since birds as well as mammals must eat often to fuel a fast metabolism, both must be efficient breadwinners, and that occasionally leads to a display of what we used to think was a human exclusive: ingenuity. No longer can we claim to be the only tool users; an ape uses a stick to get bananas and a larval ant-lion kicks sand at a struggling insect to help capture it. The archer fish shoots down insects from foliage by spitting droplets of water at them. The hooded crow robs ice fishermen of their bait by pulling up their lines. The crow walks backward from the hole in the ice, pulling the line in its beak. Then it walks forward *on the line* to keep it from slipping. Taking the next portion of the line, it repeats the process until it can seize the bait.

224247

Who will eat and who will die
— the chase decides.

Both bobcat and snowshoe hare
feel at a disadvantage in this
adrenalin-fueled foot race.
Neither likes to be caught out
in the open—nor to be browsing
or hunting in broad daylight in
the first place. The bobcat is
usually more successful when
ambushing a deer from a tree
limb or stalking a grouse from
cover to cover, getting things
over with in one good pounce
rather than all this
tiresome running.

Given an even start, the 3-
pound hare can soon pull away
from the 30-pound cat (and the
photographer reports it did on
this Colorado winter morning),
just as it often outruns the lynx,
fox, marten, fisher, and
weasel.

For the disappointed bobcat
there are always squirrels,
mice, porcupines, muskrats,
opossums, and birds about,
even in winter. As a top predator in its ecosystem, a bobcat
seldom starves.

Tool-using raises anew question of animal intelligence.

A chimpanzee (opposite) uses a twig to probe a termite nest, then draws the twig out and nibbles the termites that come with it. Chimps also wad up a leaf and use it as a sponge to get a drink. A Galapagos woodpecker finch (above) uses a cactus spine to probe for insects. The sea otter (left) dives for both the shellfish in its paws and the rock it uses as an anvil for cracking it. The Egyptian vulture (top) also makes a tool of a rock, slamming it down on an ostrich egg until the tough shell cracks, to get to the meal inside.

Is this insightful learning, or merely a mechanical repetition of a rewarding food-gathering activity? A half-century debate is currently tilting to the side of learning.

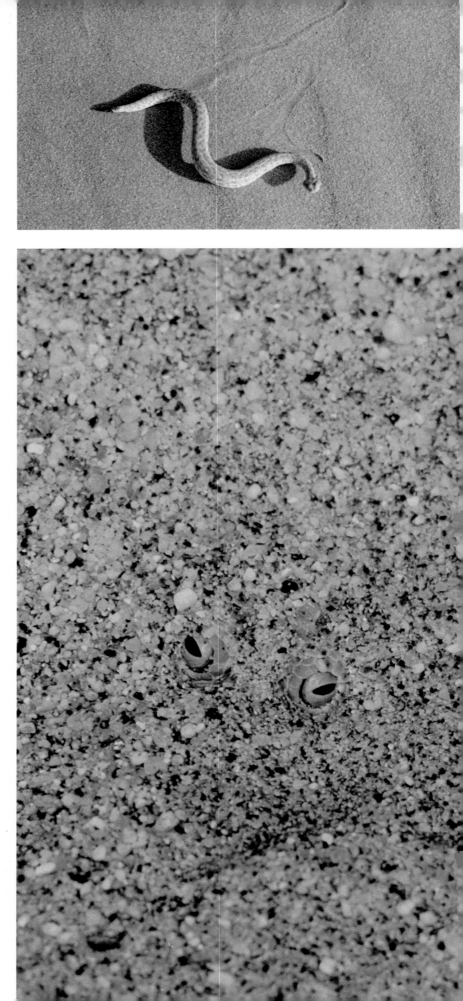

Along with eating goes the parallel need to drink, for life in all its forms needs water. But drinking serves quite different needs from eating. Water is needed not for food or energy, but as an essential partner in the body's chemical workings and as a medium for getting rid of wastes. All animals need it—from the camel that can chug-a-lug as much as 50 gallons after a long desert march, to the kangaroo rat that gets its water without drinking at all.

How does the little rat do it? Hopping about the parched and wind-scourged desert of the American Southwest, the kangaroo rat makes its own water from carbohydrates in the seeds it eats. And it hoards that water ingeniously, squandering very little in its thick urine and virtually none in its stone-hard droppings. It even uses special nasal baffles to condense the moisture from its own breath. It has no sweat glands, and so must keep cool in a burrow by day and forage by night.

That has its risks, for the silent owl rules the night, and the kangaroo rat is on its menu. By night the rattlesnake also makes its rounds; even the largest kangaroo rat will fit into that gaping mouth once its jaw is unhinged. The rat evens the odds with "tuned" hearing. Its extremely sensitive ears are especially alert to the faint scrape of snake scales on sand and the whisper of air through owl feathers. When the rat doesn't hear either in time, it serves the victor as one of *its* sources of water.

Next day it may be the rattler's turn to surrender its life—and its hard-won moisture—to a desert denizen who seems more farce than fierce, a zany bird the early settlers guffawed over and affectionately dubbed the roadrunner. Its species will probably lose the gift of flight eventually. Small loss; who needs wings when you have a long tail for a rudder and a pair of feet that take 22-inch steps 12 times a second as you sprint and zigzag at 15 miles an hour?

Bad enough that the fearsome rattler succumbs to such a comic; even more ignominious is the way it succumbs. Round the befuddled snake the roadrunner darts and dances, kicking sand in its eyes, pecking its head—and finally gulping down as much snake as its innards can hold. The rattler's tail then lolls from its beak until the other end has been digested.

eception in the sand—a cool iper lures prey with its tail.

he Peringuey's viper (left and elow) has neatly solved the roblem of finding food in the ehydrating heat of the Nanibian desert. The pencil-thin, 2-inch snake—also called a idewinding adder—simply disppears tailfirst into the sand, njoying the coolness of a 35-egree (F) drop in temperature a few inches down. Leaving only its eyes and its tail exposed, the venomous little snake waits in almost perfect camouflage. When a lizard investigates the dark, wormlike tail, the viper strikes, paralyzing its victim, then swallowing it headfirst. Lizards are a source of moisture as well as nourishment for the viper; another source of water is the dew of the cool desert night.

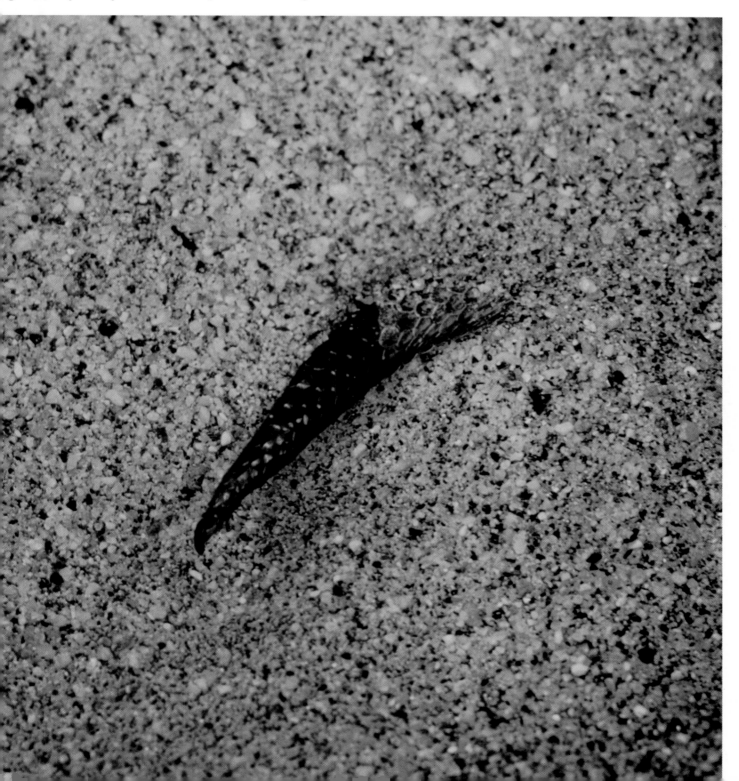

In wet worlds as in dry, the search for the next meal goes on, often in sharply contrasting ways. Gulls solve the hunger problem by eating almost anything. Some follow fishing boats crying for a handout while others scavenge on garbage landfills on the shore; further inland they follow the farmer's plow to snatch worms and grubs in the turned earth.

Far more choosy is the lesser flamingo of eastern Africa which eats nothing but blue-green algae in the Great Rift Valley lakes. Its niche is secure as long as the blue-green algae hold out, but the more finicky a bread-winner is, the more vulnerable it must be to the environment's slightest twitch. On our own shores, the Florida Everglades kite dines almost exclusively on apple snails, for which its bill is a perfect opener, but apple snails are a dwindling resource in the shrinking wetlands of that state.

From pond and ocean, diverse divers pluck food from a watery world in which they cannot remain for more than a matter of seconds. A pelican plummets into the waves, looking like a discarded umbrella as it smacks the surface with a foamy splash. A moment later, if its aim was true, it labors back into the air with a fish thrashing in the leathery pouch slung between its lower mandible edges. The anhinga—aptly nicknamed "the snakebird" for its serpentine neck—stays down longer, paddling this way and that after a darting fish. Back ashore it must spread its wings to hang its wet suit out to dry, for its feathers have gotten waterlogged because they have no oil on them.

In our western mountains a wren-like song-bird walks or flies right into a stream to feed. Bending its tiny body at an angle to the down-hill rush of the water, it darts about on the pebbly bottom to select a varied menu of snails, fish eggs, caddis worms and aquatic insects. If the water bowls it over, the ouzel, or dipper as it is also called, simply flies out of the water and circles around to try again.

Sometimes the tables are unexpectedly turned on hungry creatures of the land who go fishing. In a Florida slough a raccoon shuffles to the water's edge. No diver he, but a fondness for crayfish emboldens him to wade in, exploring first around a log lying just below the surface.

*Splash, gurgle, gulp! Diving
for fish is a pelican specialty.*

A fish (left) is about to be
scooped up into the open beak
pouch of a brown pelican.
Above the fish a huge, silvery
air bubble testifies to the force
with which the bird hit the
water from a height of 10 to
30 feet, a force that continues
to propel the pelican straight
to the target it sighted from the
air. With luck and good timing
it will come out of the water
with the fish thrashing in its
pouch. The pelican will then
spew a gallon of water from the
corners of its mouth and quickly
swallow the fish to deprive
piratical gulls of a free lunch.

   The pouch then deflates like
a balloon. This unique bag of
skin hangs from bony exten-
sions of the lower jaw which
join at the tip of the beak; they
are so flexible they can be spread
apart several inches for the
catch. (Note contrast in width
of lower beak in the two photographs.) The pelican's hidden
asset is a network of air sacs in
the bones and skin that cush-
ion its landings and make its
20-pound body so buoyant
that it pops effortlessly
to the surface.

Scavengers keep nature's cafeteria tidy 24 hours a day.

The gray squirrel (above) does it with a nibble, the dung beetle (top) with dogged toil, and the hyenas and vultures (opposite) with gluttonous protocol, but all are breadwinners who benefit the earth and its inhabitants each time they feed. Clearing the woods of antlers, burying excrement in small neat balls, stripping a carcass that would soon rot in the tropical sun is redemptive work. The lowly scavengers recycle nature's castoffs into new flesh, new bone, richer soil, and verdant new plant life.

Suddenly a great splashing, a dark shape lunging—and the raccoon becomes part of an alligator that lay in ambush looking like a log. The alligator can be an aggressive hunter on land when the occasion demands, but it's a master of ambush in the water—equal, say, to the seal that waits patiently beneath the ice floe until a fish-hungry penguin plunges in. Many a largemouth bass has also ambushed a duckling dinner it first sighted swimming overhead, and the snapping turtle varies its pond diet by snapping up a frog or even an unwary muskrat that swims too far from shore.

As each breadwinner earns its living, so each casts away its wastes. We humans turn away from the sight of flies feeding in horse manure, of dung beetles burrowing into it and forming huge pills to roll away and bury underground—and even from the sight of tiger swallowtail butterflies fluttering around a pool of urine. Yet how unrealistic is our revulsion when we come upon insect scavengers at work on our behalf as well as their own. We are indebted to these industrious sanitation crews for the constant cleansing of our environment. By nourishing their own bodies on the castoffs of others, these tiny scavengers take their place in the food chain, keeping nutrients circulating from life to life.

As all creatures share life, so all share death. Then nature calls in her loan; whatever the creature has locked up in its cellular vaults must now be returned to the natural economy. Without this final recycling, life would gradually grind to a halt, suffocated by the dead. First on the scene are the larger scavengers. Wheeling vultures ride the air, searching for the still form on the ground. One vulture spots a carcass and drops to the prize—a bonanza of a wildebeest. Others are quick to note their comrade's behavior; the first barely alights before the rest swoop in to compete for the feast. Wings thrash, beaks jab; food fights add a bizarre overtone to the *danse macabre*. The birds must gorge quickly, for four-footed rivals will not be far behind.

Hyenas come snarling in to claim their share. Flesh flies and carrion beetles also come to gorge before the bacteria and fungi from the soil take over. These microscopic laborers break down both flesh and bone for recycling through the soil back to the plants. Thus they complete the circle of life as a continuous three-way partnership wherein plants are the producers, animals are the consumers, and bacteria and fungi are the reducers.

By far the smallest in the triumvirate, the reducers' contribution is out of all proportion to their size. Without them, nutrients would remain locked up in dead tissues of both plant and animals. We think of a food chain as ending with the reducers, yet in truth it begins with them as well.

In countless forms and ways we have seen how land and water yield to all the breadwinners a share of earth's precious outpouring of food. And how all of it begins with sunlight falling on plants. On land sunshine activates the leafy legions all around us; in fresh water the lily pads and bulrushes; and in salt, the kelp and seaweeds, especially the single-celled diatoms and other microscopic specks of life. Riding the great ocean swells, algae and other plants perform about 90 percent of all the photosynthesis on our planet, and on that foundation the aquatic energy pyramids build. On sea and land the great food chains weave from vegetarian to predator to predator, ending at last with the scavengers and the reducers.

We philosophic animals survey the harsh scene and wonder why such brutality must be. In our sympathy for the prey, do we hark to some ancestral memory of far eons when our own kind fell prey to cats and canids? If true, then we must also ask whether our admiration for the predator—after whom we name our cars and teams and sporting goods—harks to our more recent triumph as predator supreme, killer of any creature we find edible and even some we do not. We have served our hitch as both the killer and the killed, but now by brain power alone we win exemption from the universal law. We eat, and almost never are we eaten. But we cannot claim exemption from our obligation to guard the habitat of all the others. Grazing or stalking, ambushing or poisoning, closing in for the kill or claiming the carcass, the role of breadwinner must be played by every wild creature that shares in the miracle of life. And for all of them, every green leaf counts.

The Defenders

Everything that lives must feed. But everything that feeds is also fed upon. And so nature's creatures stand at both ends of a challenging equation: find food until you become food. Eat until you are eaten. How long is "until"? That depends on how well the role of defender is played. There are many ways to play it—with strength, with speed, with guile, with dazzling ingenuity. But the only right way is to win, for losing doesn't include a second chance.

Perhaps the best defense is to run away from an attacker, for it cannot kill what it cannot catch. But such a race is only for the swift; plodders must face death squarely and try to survive by other means. An angry snort, a stance that promises a fight to the death— these tactics may halt the hunter and force *it* to switch to the role of defender. Most predators will pass up a meal rather than risk an injury that would hamper hunting the next time hunger stirred. And on that principle rests a common ploy with some very uncommon variations: the bluff.

The thundering charge of the heavy-footed and unpredictable black rhinoceros may or may not be a bluff, but what creature would want to stick around to find out? With a well developed sense of smell and fine-tuned hearing, the one-and-a-half ton rhino probably won't miss its target unless it wants to—in spite of weak eyesight. The rhino has been known to charge a moving vehicle and even to throw a man into the air with its menacing horn. But sometimes, perhaps when young are to be defended, this huge mammal wants only to get rid of the threat, and its charge is pure bluff.

The simplest bluff is the one in which a creature manages to look bigger than it is. The screech owl—after first trying to "disappear" by hiding its huge yellow eyes and squeezing itself small and thin—bluffs an intruder with just the opposite act. The wings and tail fan out, each feather erect and fluttering. In the blurry outline the confused predator senses a foe much larger than it had expected.

Insects are master bluffers. Most of them have to be, for they are ill armed against the beaks and teeth of enemies many times their size. A mantis might bluff a hawk by fanning out its wings to show two alarming eyespots, and thus startle the hawk into retreat. Many moths also have spots like huge staring eyes hidden on their underwings; when bothered by a bird, the upper wings are jerked open to reveal the eyes and often the startled bird flits off to seek less threatening fare.

Eyespots on the mantis and the moth enable these creatures to inject an element of surprise into a desperate situation. A bird's threatening approach can stimulate other surprises as well. At the first peck some moths will suddenly flap their wings and rock from side to side. The moth's abrupt change of behavior might cause any bird to hesitate. In the meantime this flurry of activity warms up the moth's flight muscles. Now it can take off before the bird has a chance to figure out what happened.

Some caterpillars also flaunt eyespots. Many have survived to take wing as moths simply by hunching their backs. Inconspicuous markings on either end of these caterpillars suddenly become great staring "eyes." To the stalking toad those eyes signal "snake," and so the toad swells its body and rears up its rump in bluff against a bluffer that could have been a meal. Some worms go the act one better with *two* pairs of eyespots, doubling the impact.

The eyespot effect can be just the opposite in the realm of the sea. A prominent eyespot on a fish's aft end will direct a hunter's strike toward this bogus "head"—and give the intended victim the split second it needs to zip away in the direction the real head is pointed.

In fact, the sea abounds with fakers. The stonefish looks exactly like a stone. The bottom-dwelling flounder looks just like the bottom on which it lies. A great many creatures live in plain sight of their enemies and survive by looking like something else.

The practitioners of camouflage are all around us—if we could but see them. Touch a brownish twig; it falls lifelessly to the ground and lies there. Later it crawls away, for it was not a twig but a worm. Look closely at a sprouting green leaf, and suddenly you discover a live katydid. On the patterned bark of a tree trunk hangs a moth—or perhaps many moths— in perfect mimicry of both pattern and color.

This unpredictable heavyweight may defend its territory or its calf by charging full force—or it may turn and run, crashing through dense and thorny brush at a speed of 25 to 30 miles per hour. Its thick, armor-like skin protects it against the tough vegetation, or against a lion or a pack of hyenas in the event of a fight.

The black rhino is more aggressive and less sociable than its relative, the white rhino, and is more likely to charge other animals and almost anything else that comes its way—even a disturbing sound or smell. An individual animal's reaction to man is connected with its own past experience. The calf also learns from its mother how to deal with human beings; therefore behavior varies from full flight to a charge that may or may not be merely bluff.

Touch one; if it has eyespots it will flash them at you. But if it has none it will stay put, for its best defense is still its camouflage and any movement would spoil the act. Indeed, this is the cardinal rule for creatures that hide in plain sight: don't move. Yet sooner or later all must break the rule, whether to feed or mate or carry out some other part of life's complex program. The twig crawling along a stem is a twig no longer; and the first bird to find out will be rewarded with a tasty worm.

The ptarmigan in snow can move with lesser risk than the worm because the bird's snow-white plumage blends in wherever it goes. And as the seasons change, so does its coloration—to patchy brown and white in the patchy snow and tundra of spring, then to an all-over mottling of brown when the snow has melted away. Alas for the ptarmigan, two of its foes—the arctic fox and weasel—share its habit of changing outfits to match the season. So it's as hard for the bird to spot the skulking predator as it is for the predator to see the stock-still bird.

Many creatures of land and lake are darker on their backs than on their bellies. The difference is one more aid to hiding because it tends to balance out the effect of light from the sky, which illuminates the back but leaves the belly in shadow. The caterpillar of the eyed hawk-moth doesn't know this, but if you move it to the upper side of a twig it will hasten to the underside again. Its reverse coloration—light back, dark belly—flouts the ordinary fashion, and so it instinctively hangs upside down to achieve the same effect. The caterpillar's eccentric behavior and appearance probably evolved hand in hand, leaving the defensive principle intact: dark side up, whichever side of the animal that may be.

With its naked tail, pink nose, black trim, and pallid fur, the opossum blends with few of nature's backgrounds. So it depends first on shambling to safety, then on bluff, then perhaps on fighting back—and finally on a last resort that carries to an extreme the tactic of bluffing: it plays dead. Eyes close, tongue lolls, toes curl as the animal lies motionless on its back before its confused tormentor. The attacker may nuzzle it, nip it, thrash it about,

36

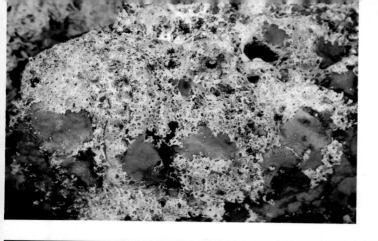

Many creatures defend themselves by hiding in plain view.

The dark back of the Colorado River toad (opposite, left) helps it to escape the notice of a hungry bird or fox. But when the camouflage fails and the toad is eaten, it has a post-humous revenge: the poison in its body produces temporary paralysis in the predator.

In Panama the sphinx moth (opposite, right) comes to rest on a surface that matches both its color and pattern. It even alights so that the direction of its pattern lines up with that of the tree bark. Many a keen-sighted bird will pass right by the stock-still moth without seeing it.

Employing the same subterfuge in tropical waters, the stonefish (top) looks exactly like the surrounding stones— including the algae covering. Thus it hides from sharks and rays that pass over it.

The long-tailed weasel (above) hunts the willow ptarmigan (left), yet they use the same defense: both switch colors to match the landscape as the seasons change. And sometimes they share a common fate: death in the claws or jaws of a hawk or fox.

even wound it with bites, but the ruse continues without a flicker. Most predators seem to need the stimulation of movement to incite them to kill. Unless they are very hungry, they will soon lose interest in an inert body, leaving behind a perfectly good meal that will eventually "come back to life" and limp away. If any creature earns the right to nine lives, it is not the cat but the opossum.

Playing 'possum is probably not a conscious act. When a hawk swoops down upon an opossum, the mammal's limited brainpower most likely cannot process the sudden logjam of high-intensity stimuli and simply shuts down, throwing it into its "act"—actually a state of shock known as thanatosis. You can induce a similar trance in many a creature simply by flipping it onto its back. A zoologist, in the course of his work, will flip an alligator; and with a gentle rub on its snowy belly may be able to keep it in that upside down position almost indefinitely.

Most creatures seem to excel at only one defense tactic. Some, like the screech owl or opossum, can bluff in more than one way. But for putting on a show with the most variety, an Oscar must go to the hognose snake.

This harmless reptile's virtuoso performance is a ruse in three acts. Act one: The Rattler. Color mimicry gives it the look of one of its dreaded cousins, and a broad head and flattened snout add to the rattlesnake illusion. Act two: Larger Than Life. When threatened, this aptly-nicknamed "puff adder" bloats to nearly twice its normal girth, and throws in a menacing hiss for added effect. Act three: The Death-bed Scene. The snake rolls on its back, tongue dangling out a corner of its open mouth. It is apparently dead. In fact, it is insistently dead, for if you roll it right side up it will promptly roll onto its back again and again.

To brainy humans, that insistence is—well, a dead giveaway. But some predators may not see it that way. The sight of a snake alive, right side up, and moving, triggers a catch-and-eat response in a skunk or mongoose, but the look of a dead one may inspire only disinterest or disdain. The snake has a last chance at survival if it can displace the former response with the latter.

A much more common defense takes advantage of the fact that many animals have a keen sense of smell. If one odor can attract a predator, then another odor can as easily repel it. The stinkbug, the stinkpot turtle, and a reeking roster of other bad-smellers survive by blunting an attacker's drive not through its eyes but through its nose.

The skunk, best known stinker among the animals, is very selective about dispensing its chemical ammunition, storing it in an anal arsenal against the hour of need. But even when that hour is at hand, the skunk seems reluctant to fire. First it raises high its striped tail and drums its forepaws on the ground in warning. The spotted skunk of the American West even does a nimble handstand. Only when its attacker ignores the warnings will the skunk whirl around and spray its evil-smelling irritant, usually targeting the eyes—and usually scoring at ranges of up to a dozen feet.

The skunk's conspicuous blacks and whites are easily seen—and easily remembered by a coyote once on the receiving end of its chemical barrage. Thus the skunks have opted not for camouflage but for bold colors and unmistakable warning. Most prowlers take the hint and give them a wide berth.

Still other creatures defend themselves by tasting terrible. And many of these have found, like the skunk, that it pays to publicize. For by the time a predator savors its terrible taste the prey may be dead; better to warn away the attacker before things reach that sorry state.

Several foul-tasting moth species have developed the uncanny ability to monitor the high-pitched "sonar" of their predator, the hunting bat; "ears" on the moths' midsections detect the bat's shrill squeaks. Some moths answer the bat with auditory signals that warn it away from a distasteful dinner. Others take evasive action. Like tiny fighter planes, attacker and defender zigzag in an aerial dogfight, darting about with dazzling agility. In a twinkling the duel stops—in one of two ways: the moth at the last split-second folds its wings and tailspins to the ground and safety—or the bat scoops up the moth in the membrane basket between its tail and rear feet and bends its head down to eat a meal it will soon regret.

Bright colors, bad smell, and terrible taste can save lives.

The vivid colors of the arrow-poison frogs (opposite) warn predatory snakes, birds, and mammals that they are dangerous to eat. In fact, the frogs' venom is so potent that South American Indians use it for their poison-tipped arrows, which are reputed to kill with a mere scratch of the skin.

Stinkbug (top left) is the nickname given to the kind of shieldbug that ejects an evil-smelling but nonlethal fluid when molested. But the bird that eats the terrible-tasting insect remembers its distinctive appearance and avoids the next one it sees.

One whiff of the stinkpot, or musk turtle (top right), is usually enough to discourage even the hungriest alligator. The foul-smelling fumes are secreted from glands on each side of the turtle's body where skin meets carapace.

The garden tiger moth (left) usually rests with its upper wings folded over its underwings. But the bat or bird that is tempted to eat this pretty creature is warned of the moth's poisonous nature by the display of its colorful underwings.

If provoked enough, the striped skunk (below) will eject its disgusting fluid into the coyote's face. But a wilderness-wise coyote will take warning from the skunk's black-and-white coloration and raised tail and give it a wide berth.

41

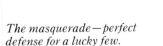

The masquerade—perfect defense for a lucky few.

With its broad hairy body, the black-and-yellow striped robber fly (top right) looks a lot like a bumblebee (top left). So bee-shy birds avoid it; and the fly can approach unsuspecting bumblebees on which it preys.

Two characteristics differentiate the poisonous eastern coral snake (opposite, top) from its imitator, the scarlet kingsnake (center)—a black snout and red and yellow stripes that touch. But predators don't notice, so the nonpoisonous kingsnake goes free.

In size, color, and pattern, the viceroy butterfly (below) is so similar to the monarch (right) that a jay, having tasted a poisonous monarch, will refuse to touch a nonpoisonous viceroy. And the ant-mimicking spider (opposite, left) probably escapes the notice of many spider-eating birds and lizards because it looks so much like the ant whose company it keeps.

The bad-tasting moth tried to warn the predator by sound. Other insects do the same by sight. You can spot a monarch butterfly a meadow away. But how can the monarch survive when it attracts the eye of every hungry bird in sight? The answer lies in the monarch's real defense: it may be poisonous. As a larva, it probably ate nothing but milkweed, including certain species carrying a toxin that makes a vertebrate retch and in heavy doses can stop the vertebrate's heart. Monarch larvae that feast on these particular milkweeds carry the poison into adulthood yet feel no ill effect themselves. But the unfortunate blue jay that eats one of these carriers retches—and remembers.

That is small comfort to the now-dead butterfly. Its flamboyant costume has indeed proved suicidal. However, nature's concern is not for the individual but for the species. A single monarch has died, but that jay will not attack another monarch, and thus the species' chance for survival has been enhanced.

But what of the monarchs that never ingested the milkweed's poison? They get a free ride, for the jay can't tell them from the loaded ones, so it shuns them all. Other butterfly species get a free ride too; those that mimic the monarch's royal raiment may flit about unmolested by predators that have learned the monarch's bitter lesson.

Such imitators abound in the ranks of the defenders. A harmless robber fly looks like a well-armed bee. A non-poisonous scarlet kingsnake imitates almost perfectly the red, black, and yellow bands of the deadly eastern coral

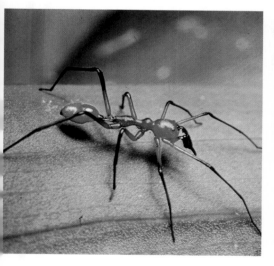

Desperate burrowers dig themselves out of sight in minutes.

The strong leg muscles that make expert diggers of the three animals shown here also make them fast runners. Given fair warning, each will sprint for its burrow when danger arises. If overtaken, each has a second line of defense. The badger (above) clacks its teeth, bites, and scratches; the armadillo (right) curls up in its shell; the sand-dwelling Coachella Valley fringe-toed lizard (opposite) hides behind anything handy. But if the predator closes in, all three turn to their ultimate security: the earth. Plunging headlong, the lizard disappears into the loose sand of its desert habitat; making the dirt fly, the other two dig for dear life—and often win.

snake. The ant spider manages to look like an ant, even to zigzagging about on six legs while it waves the other two like antennae. The wisdom of the spider's ruse is a little obscure, for in fooling hunters hungry for spiders it offers itself up to those out for ants. Possibly the masquerade is not a defense at all, but an aid in carrying out some other function of the unspiderlike arachnid's life-style. This imitation ant may even be a gate-crasher, gaining entry to the ants' underground fortresses.

Burrowing for safety is another of nature's ingenious defense mechanisms. The shrimp wallows into the sand and feels perfectly safe—until the cuttlefish happens by, squirting its water jet at the sea floor to see what it can uncover. Even if the jet blows the sand off the shrimp, the cuttlefish might easily overlook its camouflaged quarry—if the shrimp would only hold still. But the shrimp's internal programming tells it that there is safety in sand, so it hurries to bury itself again. Alas, the cuttlefish is faster. Triggered by the shrimp's movement, its tentacles shoot out and lock on with their suction pads, and the sea is suddenly poorer by one burrower.

A few talented land burrowers, caught too far from home in a desperate situation, will also put their faith in their digging prowess to save themselves. Throwing up a shower of dirt, the foot-and-a-half-long armadillo can dig itself quickly out of sight. The badger's long, strong claws on short legs on a flat body make it the perfect earthmover—and it practices a lot. In summer it may dig a new den every night. A young badger in the path of a swooping eagle may not escape its talons on the first pass, but if the eagle misses, the little burrower has a chance of digging to safety.

Many who cannot dig well themselves take over the abandoned digs of those who can. And in an emergency, any hole will do. Well, almost any; a desert mouse diving into the nearest hole to escape an owl may dive right out again with a resident tarantula at its heels.

For many burrowers home is also fortress. The woodchuck, the mole, the prairie dog and the chipmunk dig homes varying from a woodchuck's simple tunnel to the prairie dog's labyrinth with a dozen entrances.

Such a home may be proof against owl and hawk, but not against the stout-clawed forepaws of the fox or the burrow-shaped body of the badger or snake. When these invaders dig or slither in, the occupant must either dig faster than its pursuer or make for a second exit. And advance warning can give the defender a crucial head start. Thus prairie dogs post sentinels throughout their communal towns and scamper for their burrows at the lookouts' shrill cries of alarm. But they may not dive in right away; they seem to know how close they can let a coyote approach.

Behaviorists call that safety zone the flight distance. A lizard will let you approach, slowly, to perhaps a yard before it scampers. But an antelope may start running while you are still some 200 yards away. One determinant of how soon an animal will flee is the size of the pursuer in its field of vision. The bigger the pursuer's image, the sooner the animal runs, for the size of the image it sees is one clue to how close an intruder is. Thus a toddler can often approach a rabbit more closely than the child's parent can before the animal bounds away.

Among herd animals such as white-tailed deer, the sooner *all* members see one start to run from an attacking coyote the sooner they will all run, and the better their chances of escape. To this end they have evolved conspicuous features that serve to spread the alarm. A white-tailed deer blends marvelously with the dappled browns of a wood—but spook it and its tail becomes a white flag signaling danger as it bounds away. Thus alerted, others bolt for safety too, tails up.

To humans there is humor in the "pronking" of the South African springbok as it springs stiff-legged some eight feet straight up a few times before it begins to run from danger. But the purpose is deadly serious, for as each pronking antelope makes of itself a signal flag for its comrades, it also momentarily overloads an attacking lion with a confusing jumble of shapes and motions that may befuddle it just long enough to buy the herd an edge. And this edge is important. The sprinting lion must either make its kill very quickly or pant in empty frustration as it watches the more enduring springbok keep hoofing on and on.

Flight rather than fight is the speedster's first choice.

Lightning-fast impalas (below) take to the air when startled by lions or a leopard. Seemingly effortless leaps carry them high over vegetation and each other. Even when there is no obstacle to clear they take 15- to 25-foot leaps in a mad scramble toward the relative safety of savanna woods.

Like their rare springbok cousins (bottom), speed is their main defense. But the gazelle-like springboks have channeled their jumping ability into a useful reconnaissance and warning device: when alarmed, they spring stiff-legged 8 to 12 feet straight up and down several times before they take off.

The 300-pound ostrich (opposite), grazing companion of the impala and springbok, has been clocked at 30 miles per hour in its attempts to avoid becoming a lion's meal. But if cornered, this fastest and largest of birds puts those legs to another good use: kicking and slashing with heavy toenails that can slit a lion's hide.

Creatures clad in scales, quills, or shells need not run from foes.

Immobilizing a turtle which it may eat later, the tough-hided alligator (below) has only man to fear. Its eggs and young, however, when not protected by the mother's jaws, are vulnerable to swamp predators.

Retreating into their shells won't help the tropical tree snails (bottom) when they fall into the hands of their major predator—the human collector. But when a fox nudges the box turtle (opposite, bottom) this tactic works fine. Like the shell of the turtle and the snail, the overlapping scales of the coiled pangolin (opposite, top) protect a soft face and body. Twitching its sharp scales adds force to the deterrent.

The thrashing quill-filled tail of the porcupine (opposite, center) will plant its spines in an attacker's flesh. Each quill has a needlelike tip that is covered with hundreds of tiny diamond-shaped scales. The barbed quill expands as it absorbs moisture from its host.

One of the best ways to flee is to fly. The insects were the Wright brothers of the animal kingdom; for millions of years they fluttered over the heads of their earthbound pursuers. Then from the ranks of the reptiles rose the birds in their thousands of species. Today insects and birds are still the most agile escapists. The dragonfly expertly dodges its pursuer after a speedy take-off. The tiger beetle abruptly takes off and lands a short distance away after avoiding a hungry kingbird.

A ringed plover, alarmed by a snake nearing its nest, combines that grand old deception, the broken wing act, with the sublime mobility of flight to lure the snake away. First the bird drags along the ground feigning helplessness. The performance is often accompanied by cries of pain. But the plover always manages to flutter just beyond the reach of the hungry snake who is being led further and further away from the nest. Suddenly the plover makes a miraculous recovery and flies away.

No one ever flew in a suit of armor. And so the armored ones—the turtle, the alligator, the pangolin—make a different compromise; they sacrifice speed but become walking fortresses instead. Annoy a pangolin, and it will fascinate you by rolling up into a hard-shelled ball. To a formidable arsenal of teeth, claws, and thrashing tail, the alligator adds the protection of a scaly hide that shrugs off all but the most savage assault—by another alligator, usually in defense of its mate or territory. And who as a youngster hasn't run out of patience waiting for nature's little tank, the box turtle, to get up the nerve to open that all-but-impenetrable hatch and poke out a foot or a head?

The porcupine can also afford to plod, for although it isn't armored, it is definitely armed. Its deadliest array of quills are in its tail. When assaulted, it turns its back toward its antagonist and flails away. One good swat makes a pincushion of a wolf's snout or a cougar's paw. Barbs on one end of the quill dig into the enemy's flesh and pull the other end free of its sheath in the porcupine's hide. Each flex of the victim's flesh draws the needle in deeper; in time it may even reach a vital organ and cause death.

*The tail of the green anole
(left) easily breaks off when
gripped by a snake's jaws or a
bird's beak. The predator's
preoccupation with the wiggly
segment allows the anole to
escape unharmed—and begin to
grow another tail.*

*Similarly, the sea star (oppo-
site, below) can deliberately
detach an arm that is in the
grasp of a marine predator—
in fact, it may lose four of its
five arms and successfully re-
generate replacements. The
starfish pictured here seems
to have been regenerating over-
time: its new spare parts may
give it the edge in the next
encounter with a predator.*

*Regenerated parts are seldom
perfect duplicates of the lost
appendages, but the owners
seem oblivious to the change
and carry on as usual.*

A porcupine is formidable, but it isn't invin-
ible; a predator clever enough or quick enough
an flip it on its back and win a meal. So the
orcupine keeps to the safety of the trees as
much as possible.

Even with such wondrous weaponry, how-
ver, flight is better than fight, and some crea-
ures will even go to the drastic extreme of
eaving parts of their bodies behind so that the
reater part may get away. Faced with a fight,
green anole counts on its wriggling tail to at-
ract the foe's attention—then detaches it,
vriggles and all, to keep the foe occupied while
skitters to safety. A spider or crab may sacri-
ce a leg to buy its way out of a life-and-death
howdown. Miraculously, that tail and leg grow
ack again—not a perfect match, perhaps, but
n most cases as good as new.

The honeybee sacrifices more than that. One
thrust of its poison stiletto is all it gets to make
when the hive is attacked by a hungry bear.
The hollow stinger then tears free of the bee's
abdomen, poison sac and all; muscles stay with
the sac to keep pumping the venom into the
victim even after the bee has flown away. Without its weapon, the bee soon dies—but the hive
has been saved. Like the monarch butterfly,
the individual animal has lost its life but its
species will benefit.

In a society like that of the beehive, it is perhaps inadequate to say that one defending bee
has given its life for the good of the hive. For
the hive is almost a single organism; like the
spider or the anole, this organism has sacrificed
an expendable part of itself to assure the survival of the rest.

What if everything fails in a creature's repertoire of defenses? The fleet-footed deer is run to exhaustion by the pack of wolves; the wild turkey, too weak to fly after a hard winter, is cornered by a ravenous bobcat. Camouflage, mimicry, bluff, and flight — all have failed. Now the tormentor closes in.

Expect a desperate battle, for the defender is now fighting for its life. Large birds bludgeon an attacker with the hard leading edges of their wings; sharp beaks instinctively jab at the vital and vulnerable eyes. Grazers' hooves lash out like claw hammers.

But what of life at the top? Who would dare affront the hulking elephant, the unpredictable grizzly bear, or the largest creature that ever lived, the majestic blue whale? The tiniest of creatures would: the disease microbes whose home is in gut and hide and muscle. Against foes they cannot see, enemies they cannot battle, even the lords of nature come crashing to earth and never know why.

And who else would dare? We would. Defenders large and small have no defense against our looting of their habitats; their numbers shrink as our perceived interests are advanced. Perhaps their ultimate defense is not tooth or claw or stench or poison; perhaps it is the efforts of concerned people who fight to defend wildlife's right to living space.

Through the many encounters of predator and prey, nature — when left to her own devices — kept the breadwinners and defenders in a kind of evolving balance. We have permanently damaged that balance, yet nature has an amazing capacity to recoup, to survive. We must take full responsibility for ensuring that survival. There are still some islands of wilderness where each advantage developed by a hunter is answered by a counterbalance in the hunted. And this must be, for if either outstrips the other, both may perish.

Here a weasel fails to see a plump white ptarmigan in the snow; there a dragonfly snares a fast-moving mosquito in its basket of legs; somewhere an eager fox pup grabs a toad and can't spit out the foul-tasting thing fast enough. Over the uncountable centuries, each victory, each defeat will continue to add its tiny grain to this ever-changing balance.

When all else fails, predator and prey fight to the death.

Only a perfectly executed ambush could enable a lioness to pull down a zebra without giving chase first, as this one did at an African waterhole (opposite). One leap and her claws were in the zebra's shoulders and her fangs in its neck. The zebra had no chance to use the speedy legs nor the sharp hooves which had saved it so many times before.

The cornered cobra (above), like the zebra, is fated to be the loser in an unequal battle. The agile mongoose teases the snake into striking several times, always eluding its venomous fangs. As the cobra tires, the quicker mammal pounces and seizes the back of the hooded head while it is down.

Victory enables the lioness and the mongoose to fight again — until a predator gets the mongoose, but only age or disease will defeat the lioness.

Seeking a Place

To find food and a
mate many animals
become travelers, and to
shelter their young
many become builders
or squatters.

The Travelers

A wandering albatross unfurls the longest wingspan of any bird—nearly 12 feet tip-to-tip—and rides the sea winds for perhaps a million miles before its span of years ends. A bobwhite quail hatches in a midwestern meadow and almost never ventures more than ten miles from its birthplace. A barnacle glues itself to a steel piling so permanently that some of the steel will come with it if it is scraped off; barring that, it completes its adult life without moving the breadth of a hair.

There is not one animal that does not travel from one place to another at some point in its life, by some means, for some reason. Even the barnacle spends its several larval stages as a free-swimming speck called a nauplius—and as a stuck-fast adult it may still span an ocean on the hull of a ship or the flank of a whale.

There are many reasons to travel. Most breadwinners travel to find food or to catch it; most defenders travel to get away from enemies as fast as wing, foot, or fin will propel them. By traveling, a mate may be found, a family raised, a cold winter avoided.

Sometimes an entire animal population will take off because the food has run out. This can happen when a plentiful food supply has produced several favorable breeding seasons in a row for a particular species; then the food supply drops, bringing the threat of imminent starvation. When that happens to snowy owls or waxwings, flocks of them suddenly leave the overpopulated area and travel to places far beyond their normal range. This spontaneous irregular movement is called an irruption.

When the brown lemming community becomes overcrowded, they become frenzied and hysterical. Thousands of these small mammals scramble out of their burrows in the tundra and fall all over each other as they rush blindly away. Some drown in rivers or lakes, some run right into the mouths of predators. But a few stay behind and survive; they ensure the future of the species. And for several years there will be no more irruptions.

For some creatures, travel is so much a way of life that no one place can be called home. These are the nomads, the species that wander over an unpredictable route in their search for food. Both Indian and African elephants are nomadic, and so are the army ants of Africa and the American tropics. The nomad lives by a simple rule: when an area has been depleted, it's time to move on to a better place.

Emigration involves moving out of an old home into a new permanent one. Normally sedentary wild rabbits in Australia, when faced with a drought, travel far in their search for food and water. But once they have emigrated to a spot where their needs are met, they settle down and become sedentary again.

But to us the most fascinating travelers of them all are the migrants—those that make one or more predictable round trips within their lifetimes from a breeding area to some more-or-less distant place, and then back to their birthplace again. How do snow geese in their thousands wing their way a spectacular 3,000 miles each autumn, then flock home again in the skies of a North American spring? What calls the incredible little barn swallows across the thousands of miles between the tropics and the north? How are these restless yet purposeful journeys so perfectly navigated? Man has yet to find the answers to these questions. From light plane to jetliner, we cram our aircraft with navigational aids and still there is scarcely a pilot aloft who has not at some time felt the lonely chill of being lost. Yet bird and butterfly, whale and caribou all make their ancestral journeys with no "instruments" except what they carry inside them.

In the air, on the land, and through the water the travelers make their way. Even a land-lubbing spider may find itself making a sea voyage. Ships many miles at sea find tiny spiders dropping in on them from the passing breeze. Somewhere on the far-off shore the little aeronauts, a few weeks out of the egg sac, seem to answer an inborn urge to emigrate to a new territory. Beckoned by a helpful wind, each young arachnid climbs a stalk or fence-post, rears its abdomen into the air and squirts out a long strand of silk. Yanked aloft by the breeze, silk and spider surrender their destiny to the whim of the wind. It is appropriate that this ingenious but risky mode of travel is called ballooning, for like the balloons we humans contrive, there is no way to steer.

Clouds of migrating snow geese mark the changing seasons.

After weeks of riding the skies at altitudes above 2,000 feet, lesser snow geese touch down on a wildlife refuge in New Mexico. In quest of their winter diet of grasses and grains, tens of thousands of the most abundant of North American geese lift off from their Arctic breeding grounds beginning in mid-August. Chattering shrilly as they go, the snow geese fly swiftly in long diagonal lines or in V-shaped flocks. Rest stops, lasting from a few days to a few weeks, lengthen the trip, the stragglers finally showing up in late November.

When the spring breeding time urges the snow geese north again, they are in a much greater hurry. With 22 days required for incubation of the eggs, and another 42 days until the young are ready for the fall flight, there is virtually no slack in the time table. By late May all have arrived at America's northernmost nesting grounds.

Previous page: Olive Ridley Turtles—Pacific coast, Costa Rica.

Some of the spiderlets may travel only a few inches if the wind quits. But others could conceivably cross oceans. Some experts theorize that certain species have spread from one continent to another by ballooning. And that's quite a trip for a sedentary creature that traps flying insects but never grows wings of its own.

The winged insects also accomplish remarkable journeys. Dragonflies are hardy enough to fly hundreds of miles. So are ladybird beetles—much to the dismay of the gardener who buys them from distant suppliers, sprinkles them on aphid-infested flowerbeds, and sometimes discovers in a few days that "Ladybug, ladybug, fly away home" is not just a nursery rhyme.

Few insect travelers chill the marrow of man as do the short-horned grasshoppers we commonly call locusts. In the savannas of Asia and Africa these insects usually live as loners, each munching its helping of an abundant food supply. But sporadically, responding to the promptings of an internal clock or a violent change in the environment, or perhaps to both, the locusts begin to collect in groups. Mating begins with the onset of the rainy season, and the group becomes a swarm. If food becomes scarce, the large and normally sedentary population is forced to migrate.

Swirling into a menacing dark cloud, the swarm begins to move across the land. Borne by papery wings on a good stiff wind, each locust flutters to the swarm's leading edge, then drops to feed on whatever grows beneath. Ten billion locusts may buzz overhead as the feeder chews until the swarm has nearly passed. Then it leaps aloft again to fly to the front

Wings bear migrant insects afar, wingless spiders ride the wind.

As millions of desert locusts (opposite) devour all the vegetation in an Ethiopian valley, they also destroy themselves — dying as the food runs out.

Ladybirds (top) also migrate in search of food, but enjoy a more stable life. They shuttle hundreds of miles between their hibernation spots in the mountains and feeding grounds in the lowlands.

Seeking a home rather than food, ballooning spiders (above) entrust their destination to the vagaries of the wind.

*Monarchs gear journey to south-
ern flowers, northern milkweed.*

The monarch's flight, traced by
researchers named on the tag
below, begins in leisurely
fashion at summer's end. To
escape the northern winter, butterflies just out of the chrys-
alis glide and drift 2,000 miles,
staying 15 to 20 feet above the
nectar-producing flowers on
which they stop to feed. By the
end of October, millions cling to
slender-leafed trees (opposite)
in Southern California, Mexico,
and along the Gulf Coast. Millions more winter in warmer
parts of the same areas as
free-flying individuals.

Each day the roosting
monarch opens its wings to
be warmed by the morning sun
until at 55° F. it can take
off to feed on the nectar-
laden blossoms nearby.

In early spring urgently
beating wings carry the monarchs back north. They take no
detours and rarely pause to
sip nectar. Their tattered wings
(bottom) are witness to the
hardships of the journey they
have endured to reach the
milkweed on which their
larvae will be nourished.

and repeat the cycle. The terrible armada may
blanket 400 square miles, obliterating precious
crops in its path.

Humans dread the locust's coming, for it
brings devastation and want. But we celebrate
the monarch butterfly's migratory flight, for
it brings us beauty and wonder. Every autumn
this fragile flash of orange and black skitters
southward across the American sky, bound
from as far north as Canada to as far south as
Mexico. Great flocks of them alight in Pacific
Grove, California, where with hoopla and pa-
rades the citizens reaffirm the town's nickname,
"Butterfly Town, U.S.A." Each year the same
trees and bushes bow under the weight of
them, though each monarch weighs but a hundredth of an ounce.

It is amazing enough that this snippet of gossamer can make such a journey. What astounds
thoughtful observers is that the monarchs
spangling this tree or that bush have never
been here before; last year's resting hordes were
their forebears! When spring beckons them
northward, some of these new ones will flutter
away to the northern states and on into Canada,
there to lay their eggs on the waiting milkweed
and die. And at summer's end *their* offspring
will fly back to the same California roosts.

The "why" of the monarch's migration may
be explained by the milkweed, its principal
food source. Perhaps the insects journey northward to intercept the plants as they emerge
from winter's grip; that would oblige the new
generation to reverse the itinerary as the
plants die in the fall in order to intercept the
milkweed again the following spring. But the
"how"? What inborn map guides young monarchs to a specific food supply thousands of
miles into the unknown? We have no answer.

Many kinds of butterflies travel. And some,
like the monarch, are migrants. In tropical
Africa, near the Red Sea, the painted lady
emerges from its larval stage in March. It
reaches the Mediterranean by April, and the
north of France by May. Along the way there
may be a second breeding. And in some years
both generations cross the English Channel
to spend the summer in the British Isles. On
their return trip they follow the thistles and
other food sources back to Africa.

Fish endure tremendous hazards to spawn at their birthplace.

Tens of thousands of grunion (opposite) reach the California shore on the crest of a wave at a high spring tide. The six-inch smeltlike fish land high up on the beach where females dig into the sand, tail first, to lay eggs. Males encircle them, releasing sperm (opposite, top). But the shoreline visit is brief; the next wave carries the grunion back out to sea.
Eggs follow at the next spring tide 14 days later, hatching within three minutes after being engulfed by a high wave.
Migration is not so easy for the Pacific salmon (left). Those that don't end up in the stomach of a bear or wolf reach their spawning grounds only after a staggering journey, sometimes thousands of miles long. Ascending the stream of their birth, they may climb two thousand feet or more in altitude. Then, having survived the grueling trip, the exhausted fish spawn and die (top).

There are reasons to travel besides food. A foamy wave tumbles onto a California beach, flinging its spume at the watching moon. As the sea drags the wave back, a wiggling cargo stays behind. These are grunion, seemingly courageous little fish with a built-in urge to risk the perils of land in order to lay their eggs in its protective, sandy embrace. Another wave whisks the grunion back out to sea. Fourteen nights later, yet another wave will sweep back onto the beach and gather up their hatchlings, welcoming into the sea-home a new generation of grunion; in their time they too will come back to the beach, riding another wave ashore.

Far away in the mid-Atlantic, tiny Ascension Island prints its dot on the charts of mariners. Here green sea turtles gather offshore to mate.

Females lumber ashore to lay their eggs. They have come back to their birthplace from Brazilian waters some 1,400 miles away. There they live—but here they breed.

What force goads creatures of land and sea to seek out their own birthplace and there sow the seeds of another generation? We cannot answer the question, but we do know that in some species the pull of ancestral birthing grounds can be one of the most compelling urges in all of the natural world. And in no other species is it more impressive than in the Pacific salmon.

Hungry bears wait in river shallows to scoop the numerous fish onto the banks. Wolves hiding among the grasses grab the salmon as the hardy migrants swim past. Towering man-made dams block their route upstream, forcing

*World's largest animal fasts in
tropics, feasts in polar waters.*

*Thousands of watery miles link
tropical winter breeding
grounds of the blue whale with
its Antarctic summer feeding
areas. Ending their breeding
fast, they eat tons of tiny
shrimplike krill. Forced to the
surface by polar storms, krill
are filtered through baleen in
the whale's cavernous mouth.*

them to leap up stairlike "fish ladders" one
water-filled step at a time. Nature arranges
another obstacle course in the innumerable
shallows, rapids, and waterfalls punctuating
many of the rivers veining the northern reaches
of the American continent. Man-made pollu-
tion, hunger, low water, accident exact a terrible toll. Yet salmon return from the sea to the
streams of their birth to spawn, and only death
stills their drive to do so.

For the Pacific salmon this is life's final act,
played out in bright nuptial coloration after
two to four years of circuiting the sea in
drab array. Atlantic salmon, however, can survive to swim back downstream. And after
spending a year in the open sea they go through
the entire process of spawning again. For these
incredibly hardy fish the cycle may be repeated up to eight times.

To respond to the homing urge, a traveler
must be able to find its way to its own particular spot on the vast expanse of the planet's face.
It must know how to get there, and know when
it *is* there. In short, it must navigate. It has
been suggested that the green turtle finds
Ascension Island by homing in on its "scent,"
a chemical signature that is wafted to Brazilian waters by prevailing currents. But more
recent experimentation has led scientists to
the hypothesis that green turtles orient themselves by the sun and not by scent at all.

Can a fish remember a particular odor after
several years at sea? Apparently a salmon can,
as it returns unerringly into the mouth of its
own natal stream, and may even track a maze
of forks and channels to the same riffle where
it hatched long before. It is certainly well
equipped for such complex and intricate path-
finding. Chemical sensors in its nostrils can
detect traces as small as one part in a billion.

No such memories guide the tiny, leaflike
eels that squirm from their eggs anywhere from
330 feet to 6,000 feet down in the Atlantic
weed patch called the Sargasso Sea. These
larval wigglers ride the Gulf Stream to the
coasts of North America and Europe. There
they find streams they have never known before, and against the unfamiliar current they
ply their way inland. Some go all the way to
the headwaters, often traveling in spectacular

schools that string out for miles. It takes the American eels one year to reach their destination. European eels make a three-year trip. During their travels the larvae elongate; when they reach their new home waters they will mature into the familiar freshwater eels.

About six years later, like salmon in reverse, the senior eels quit the rivers and make for the open Atlantic, if necessary even slithering over wet land barriers in order to reach a channel to the sea. Their life and travel end in the Sargasso deeps—where new life soon begins in the legacy of eggs they leave behind.

For the European eel, travel in a watery world takes relatively little effort. Its oceanic round trip follows the track of the Atlantic's clockwise currents. It is these currents which also enable earth's largest mammals, the whales, to cover great itineraries to calve in warmer seas and then to feed and fatten in food-rich waters nearer the poles. The food that attracts the whales is plankton or krill, tiny organisms that are found in concentrations about one hundred times greater in the Antarctic than in equatorial oceans. When the Antarctic winter sets in and ice floes prevent humpback whales and blue whales from inhabiting the coastal areas, they migrate to the tropics. There they give birth to their young and suckle them, but for several months eat almost nothing themselves. As spring approaches in the polar region, whales again travel. Buoyed up by the currents, they move in search of the krill-rich polar waters.

But the whales' land-bound kin are given no such traveling aid. Weighed down by gravity and slowed by the vagaries of weather and terrain, the mammals of land must use up a lot of energy and time as well. For these reasons the migrations of land animals cannot be compared with the myriad comings and goings of seals and sardines, of the Pacific bonito and mackerel or of separate herring populations, each traveling to its own feeding grounds.

Still, zebra and reindeer awe us with their thousands upon thousands as they draw a living blanket slowly across plain and tundra. Each year hundreds of thousands of wildebeests move over the Serengeti Plain looking for calving grounds where they can go about the busi-

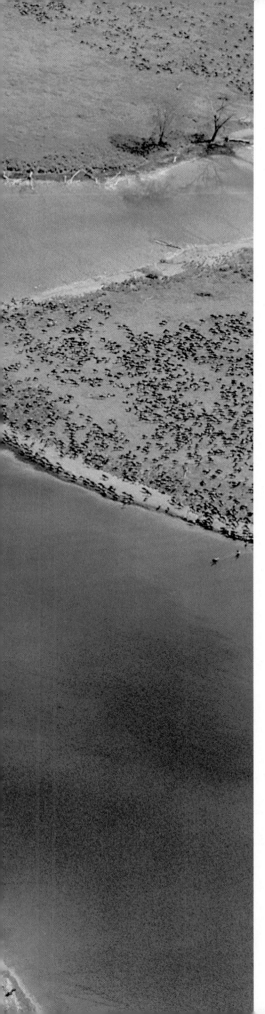

ness of bringing the young into the world in a place where there is plenty of food and water. They do not breed in the same place every year. Always searching for the succulent grasses of fresh pastures, they travel about 200 miles a season, many dying along the way from disease, starvation, and predation. African antelope, too, travel from one place to another looking for food and water. But these are more nomadic than true migrations since movement varies according to the abundance of vegetation and climatic conditions at the time. If there is enough food and water in a given year there may be little or no travel.

Caribou and wapiti are true migrators. Individual herds of Alaskan caribou may cover a distance of several hundred miles a year. Those that spend their summer on the Arctic coast find their way southward to the Brooks Range in northern Alaska as winter approaches. For other herds, summer feeding grounds are on the northern coast of Canada. Canadian caribou may travel southward as far as Ontario as the harsh winter unfolds. In midsummer they live on sedges, willows, and birch; in winter they switch to lichens, mosses, and grasses.

Throughout history there have been communities of human beings tied to the journeys of their fellow mammals. Driven by necessity to follow the migrations of reindeer across great expanses of land, the Laplander adopted a nomadic lifestyle provisioned by the reindeer's milk, meat, and hides. In similar fashion the Masai herdsman of Africa has linked his life to that of his cattle. Even today, dwindling tribes follow their animals as they eat what one spot offers and then trek on to another.

One family of migrating mammals has mastered the art of flight: the bats. Flapping their leathery membranes they shuttle from summer roosts in the temperate zones to winter shelters in the tropics. The fruit bat can even follow the ripening of its favorite foods.

But what of those small mammals whom nature has not equipped either to flee from winter's cold and want with the migrators, or to stand and face the elements with the larger animals? They must go somewhere, so they scurry to the only haven they know—their burrows and dens. And there for some of them un-

A swim across Ndutu Lake climaxes wildebeest migration.

On their northwesterly trip to calving grounds in Tanzania, just before the rains begin in early spring, wildebeest cross a large lake at the edge of Serengeti Park. Frequently traveling in single file, the gregarious bulls, and females heavy with unborn calves, hurry in search of the grass that makes up 98 percent of their diet.

With no apparent social hierarchy, they scatter during the rainy season, and converge in herds numbering in the tens of thousands when the dry season diminishes both food and water. Wildebeest can get along without water for up to five days, but prefer to drink daily.

folds an annual miracle called hibernation.

As the days grow colder and shorter, the woodchuck builds up food reserves in the form of body fat. Replete and roly-poly, it then retires into its burrow, plugs up the entrances, and curls up for a long winter's slumber. Gradually the little sleeper's temperature drops. Its heart slows to perhaps eight beats a minute. Its respiration nearly stops. But the woodchuck is alive, and even stirs occasionally, venturing up to the chill surface for a bleary-eyed look around, and a nibble if food happens to appear nearby.

Most of the hibernators are insects and reptiles and amphibians, creatures known as ectotherms because they cannot generate their own body heat. When all outside is stiffening in the grip of autumn, some turtles and frogs must burrow into the pond bottom or the forest floor to sleep off the winter months.

But the woodchuck is an endotherm; it *can* keep itself warm from within. Why, then, does it slow near unto death while some of its fellow endotherms stay active despite the cold? One likely answer is that the woodchuck hibernates to wait out the lean months in its food supply. Other endotherms eat food they can find in winter. For the deer there are cedar and balsam fir and sagebrush. Cottontail rabbits eat the twigs of sapling trees or shrubs. Hibernation then is simply one of the methods for getting through the winter months. As the March sun warms its den, the woodchuck emerges—at first just briefly during the warmest parts of the day. Gradually its summer schedule is resumed.

Of all the migrations we earthbound humans are privileged to observe, surely the most awesome and wondrous is that of the birds. We stand rooted to the ground and gaze up at the wheeling, soaring flocks. Ah, to be free as a bird! Our envy may be misplaced, for studies show that birds are not nearly as free as they seem. Of 660 species that nest in North America, two out of three take wing and move on when breeding season ends because they are compelled to by habits that were fixed in their very structure far back in the reaches of time. Bird migration is essential in nature's ongoing balancing act. But to a single European crane

tay-at-home animals seek
efuge in caves or dens.

---

he woodchuck (opposite, top),
eep in hibernation, is not
roused even when a bulldozer
xposes its den. The sound
leep that helps this fat, fur-
lad herbivore evade the
old may lead to its death—
eavy rains sometimes drown
hibernating woodchucks
in their burrows.

Little brown myotis (opposite,
middle) are among the bat species that hibernate rather than
migrate. But they awake from
time to time to flutter around
the cave, lick moisture from
its walls, and perhaps to mate.

Prairie rattlesnakes (op-
posite, bottom) may den straight
through the winter. The narrow entrances to their retreats
usually make it impossible for
other animals to get inside.

The chipmunk (below) was
found hibernating in a pile of
firewood. Though it may sleep
deeply through a severe cold
spell, the chipmunk is not a
typical deep hibernator.

*Delicate looking but strong,
these accomplished flyers (opposite) see more sunshine than
any creature in the world—24
hours of daylight for about
eight months of the year, and
more daylight than darkness
during the remaining four
months. Their odyssey transports them from their breeding
grounds—northern Greenland
to Cape Cod on the Atlantic
coast, and to the northern
British Isles and Norway in
Europe—to areas south of the
Indian Ocean, Australia, and
finally, the antarctic seas
where they winter.*

*In June, after a 21-day in-
cubation, chicks (top) hatch on
Machias Seal Island at the
mouth of the Bay of Fundy.
Parents are fearless in defense
of the eggs and young in their
hastily constructed nests,
remaining attached to the
island colony until time for
the autumn flight.*

or an Asian wheatear or an American song-
bird it can be a time of great hardship and risk.

The ruby-throated hummingbird of the eastern states flies over land until it reaches the
shores of the Gulf of Mexico. Then it strikes
out straight for Yucatan, some 500 miles, 25
hours, and more than four *million* wingbeats
away. No wonder many drop to the beaches so
spent the natives there can pick them up in
their hands; no wonder it's common to find the
remains of these and many other overwater
migrants in the stomachs of fishes at sea.

But the migratory birds that have touched
the lives of humans most directly are the waterfowl. Ducks and geese and swans have given
man food and sport and inspiration for his
artistic talents. Most waterfowl spend their
lives as members of a flock. Prodigious distances pass beneath their wings each spring
and fall. And formation flying is one of the
marvels that make such feats possible.

In V formations, an up-current generated
by the beating wing of a bird can be utilized
by the bird behind--that is, if the second bird
positions itself at the correct distance to gain
the support of these eddies of air. The lead bird,
of course, has no such help, and so it falls back
after a while to rest in a position closer to the
rear of the line. Because only one wing at a
time can be supported by the eddies, a bird
will switch from one leg of the V to the other
from time to time to alternate the wing that gets
the support. Canada geese, whistling swans,
and snow geese travel thousands of miles in
this formation, keeping up their flight, with
few rest stops, continuously, night and day.

Bobolinks of the Canadian midlands may
winter on the Argentine pampas 6,000 miles
away. From North Atlantic to South Atlantic,
the greater shearwater commutes across 8,000
miles of empty sea to find Tristan da Cunha
Island and a winter home. But none can out-
distance the Arctic tern.

Perhaps the most spectacular traveler of
them all, this hardy tern rears its broods in
the far north—sometimes less than 500 miles
from the North Pole. A change of season finds
almost every Arctic tern near the *other* pole. A
single year's migration puts about 25,000 miles
on the bird's airframe.

In evolving the birds, nature has devised
some superb flying machines. Teeth are absent
eliminating unnecessary weight. Bones are
hollow for lightness. A large keel juts out o
the breastbone; to it are anchored large, power
ful muscles that pull the wings downward in
the power strokes of flight. And on those wings
is mounted the strongest structure for its
weight that nature has yet devised: the feather

We see feathers in the fossils of Archeop
teryx, a reptilian proto-bird of 150 million year
ago. Yet this creature may never have flown; it
feathers may simply have served as nets to
snare insects as it ran along the ground. Bu
darting about with "wings" outstretched mus
now and then have lofted the creature into
an unexpected glide. And from that simple glide
have come the effortless circling of the hawk
the fearsome stoop of the peregrine, the "flight"
of the flightless penguin through water instead
of air, and the capacity that many birds have
to travel over incredibly long distances.

Birds don't migrate just because they *can*
there has to be an advantage or the habit woul
not prevail. Flying south in autumn spares the
bird from winter's cold—but that is not rea
son enough, since the feather that gives i
flight is also a superb insulator, and with i
many birds withstand the harshest of climes

The explanation becomes more complet
when we consider what happens to their foo
supply in winter. Insects vanish. Seeds ru
short. Prey animals hole up or tunnel under th
snow. If every bird stayed, many would starve
But if enough birds leave, the winter environ
ment can carry the rest through until spring.

Migration demands preparation. Fat begin
to build up in lumps, each a little storehous
of energy to power the wingbeats ahead. Th
bird becomes nervous, restless, fidgety. I
many species these changes happen long be
fore the air turns nippy. Day length is probabl
the triggering stimulus—and it may act direct
ly on the pituitary gland, a control center nea
the base of the brain. Some scientists are ex
ploring the possibility that through a bird'
thin skull the pituitary may be able to "see"
light outside and react to its duration, flood
ing the bird's system with "orders" in the form
of hormones that ready it for the journey.

Ducks and geese use north-south flyways, albatross flies west.

After arctic breeding and parenting speeded by a round-the-clock summer sun, honking Canada geese (opposite and top) migrate to the southern states.

Flocks of mallards (left), close relatives of ducks worldwide, may be seen along all of America's north-south flyways.

In contrast the lone, long-lived wandering albatross (middle), after five to ten years of learning to navigate, roams southern oceans between island breeding periods. This spectacular glider rides westerly winds for hours without flapping its wings.

12

14

Migrations follow rise and fall of temperature and food supply.

Animals on the move, portrayed by a few representative species on the map at left, weave a vast network of trails, flyways, and sea lanes around the globe, making it possible for every environment to consistently support life to capacity.

Sea creatures like the eel, green turtle, whale, and salmon astound us by their inclination to breed thousands of miles away from the spot where they find nourishment and security throughout the remainder of their life cycle.

But mountains, deserts, rivers, and forests fill potential migration paths of land mammals with obstacles. Heavy-bodied and unable to defy gravity by traveling through air or water, the caribou's several-hundred-mile journey is one of the greatest land migrations.

Birds cover long distances while expending less energy per mile—insect-eaters such as the wheatear being forced to migrate farther than seed-eaters that can find food even in winter.

1. Pacific Salmon; 2. Lesser Snow Goose; 3. Monarch Butterfly; 4. Caribou; 5. Eel; 6. Green Turtle; 7. Reindeer; 8. European Crane; 9. Zebra; 10. Wildebeest; 11. Wheatear; 12. Bonito; 13. Humpback Whale; 14. Blue Whale.

The blackpoll warbler flitting frenetically in a seaside thicket in Connecticut is fully prepared for its 2,300-mile flight over open sea to South America. Now it must choose the right time for takeoff. Like many migrants, it is an adept weather forecaster, waiting impatiently for a cold front from the north that will be followed by clear skies and a helping wind. Here comes the front—and there goes the blackpoll with a maximum chance of success.

Migration's greatest single mystery is this: how do birds navigate? Ingenious experiments have found some answers—and raised new questions in the process. We have found, for example, that some birds seem to steer by the sun. But the sun is a moving reference point and to steer by it requires a fairly precise sense of time—an internal clock whose mechanism we have yet to discover. Stars, constellations, the invisible lines inscribed by earth's magnetic field, the pattern of landmarks below—each of these may serve a bird in flight.

Prevailing winds are a critical element for small birds migrating to South America. The birds persist in their southeasterly flight despite Caribbean tradewinds blowing them toward the southwest. And so the birds land right on the mark at their South American destination. We know this because radio tracking systems have enabled scientists to follow the birds' long-distance migrations. A biologist traps a bird, attaches a transmitter to it, then lets it go. Signals indicating the traveler's whereabouts are picked up by receivers.

But the birds may have other ways to navigate that we have not yet considered. It is possible that some species hear the distinctive growl of wind over mountain ranges and the rumble of surf from shorelines, as waves cause the air to vibrate, hundreds, even thousands, of miles away. Humans cannot hear such "infrasound," for it pulses at frequencies many octaves below our bottom threshold.

Over that mountain, under that ocean, and through the air that blankets them both, the travelers of earth traffic to and fro. They are nature's movable counterweights, balancing out the fluctuations of her myriad places to live. And for their ceaseless journeyings, the travelers are equipped wondrously well.

The Builders

A hiker's dusty boot scuffs the sand of a New Mexico desert. The hiker pauses, stoops, and picks up what appears to be a corncob turned to stone. The reality of the odd object is even odder than the hiker imagines, for it tells the trained eye that this desert, now a mile above sea level, was once the floor of a shallow, shimmering sea. The "cob" is a fossil, a cast of the burrow left by an ancient shrimp that dug into the sandy bottom and made itself a home.

Far to the north in the American prairieland, a paleontologist puzzles over a "devil's corkscrew," another fossil that was once a home. This weird spiral of stone preserves the downward-twisting burrow of a long-extinct creature of the land, perhaps the beaverlike rodent known to science as *Steneofiber*.

In all of earth's environments, through all of evolution's years, animal builders have been at work. Whether they are stay-at-homes or travelers, these myriad creatures have changed—become specialized—to the point where nature unimproved is not enough. Whether to hide, to feed, to escape from heat or cold, or to rear their young in safety, they must modify their environment to minimize its perils. And so they build.

Some build by combining materials—weaving stick to stick, daubing mud upon mud—each species laboring until its inborn schematic has once again been realized. Others build by removing materials—taking dirt out of the ground or sand out of the sea floor or wood out of the trunk of a tree—in order to hollow out a home. Still others build from the stuff of their own bodies—the spider its web, the spitbug its bower of bubbles, the termite its tower of dirt hardened with its bodily secretions. And when the builders move out, the squatters move in, for in the economy of the wild, little is wasted.

To the human eye, the most conspicuous of the builders are the birds. We see their handiwork all about us—the knothole jammed with twigs and straws; the earthen riverbank pocked with nest holes; the roof overhang adorned with a cup of clay; the treetop with a fright wig of branches, feathers, bits of rope and rag and whatever else caught the builder's eye. To the casual observer, a nest is a nest; except for size, the robin's architecture may seem much the same as the eagle's. Yet so distinctive are the design, construction, and materials used by a given species that almost anyone can learn to identify a nest without ever seeing the builder.

The robin stiffens its nest with just the right consistency of mud. This cements the grassy bowl so firmly that it remains intact if you should lift it out of the crotch where the robin has wedged it. Given a fan palm, the hooded oriole may punch holes in one of its fronds, then stick palm fibers through the holes and weave them into a hanging purse that uses the frond as a roof. The burrowing owl may line its nest with dried cow dung, the kingfisher with partly digested fish bones regurgitated from its own crop, the roadrunner with snakeskin. In each case, the nest materials tell you something about the builder: this one eats fish, that one frequents pasturelands.

The size of the nesting cavity usually tells you the size of the builder, for most birds use their own bodies as template. Watch the tiny hummingbird sit and turn round and round to shape its walnut-sized nursery of plant fuzz and lichens and gossamer spider silk. Watch the bustling wren give first priority to the size of the openings it investigates. A hole about an inch across will let it squeeze in but keep most of its enemies out. Watch the weaverbird of South Africa plait a hanging wreath of grasses, then perch at the bottom of its arc and expand it into a woven sphere; how big the sphere will be depends on how far the bird can reach.

Closer to home, we can watch another master weaver, the northern (or Baltimore) oriole, and wonder why the female goes to the trouble of weaving such an elaborate hanging nursery year after year. Around a spray of branchlets she anchors the first strands of plant fiber or kite string or whatever the locale provides. Her beak thrusts and pulls as quickly as the deft fingers of a seamstress, yet it takes her about three days just to finish the framework. Finally she plaits the hanging pouch, sometimes leaving an entry at the side as well as the top, and lines the brood chamber with soft animal hair or fine snippets of grass. Let the breezes blow; her babies' cradle will sway in safety.

Tropical termite mounds rival work of human builders.

A cluster of termitariums in Australia resembles a sculptured tableau of hooded monks in wrinkled robes of brown. Each mound is indeed a sculpture, the monumental achievement of its thousands of tiny makers and residents.

Termites build the 20-foot-tall structures by mixing their bodily secretions with grains of soil, one tiny blob at a time. The mud, called carton, dries to a rock-like hardness. Walls may be built to a thickness of a foot or more. The interiors are laced with corridors, rooms, and ventilation shafts.

Far down in the cool, dark galleries the winged insects tend their queen; from her large body pour the future workers of a society that functions almost as a single animal.

The mounds pictured here are the work of sophisticated tropical cousins of the primitive wood-chewing termites that worry human homebuilders.

1

3

4

Birds weave nature's leftovers into a variety of nests.

The bald eagle's eyrie (9) and the cup of the black-chinned hummingbird (6) are among the largest and the smallest of all the nests woven by tree-nesting birds. The eagle assembles branches by the ton, the hummer gathers plant down by the gram—of softest milkweed, thistle, sycamore, and willow. Eagles may use a nest for ten years or longer, the hummers craft anew each spring.

Such contrasts abound in the work of bird weavers. Most species rely on nature's infinite variety of plant fibers. Plant stems pin the nest of the blue-gray gnatcatcher (3) together and lichens add a jewel-like touch. Grasses, rootlets, and twigs cover and line the mud foundation of the American robin's nest (8). Others, like the chestnut-sided warbler (7), may add man-made materials. The gnatcatcher, the hummer, and Australia's red-capped robin (4) bind their nests with spider's silk.

Nor do the weavers agree on such architectural details as the location of the door. The cactus wren (1) enters at the end of its 18-inch tunnel in the arms of a saguaro; the winter wren (2) on the side of its ball of bracken; and the masked weaver (5) from the underside of its pair of hanging nests.

5

6

7

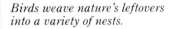

8

*The hole nest—a home
built of nothing but work.*

*A family of burrowing owls
(below) enjoys the peace and
warmth of a home in the earth;
yet, even here, danger lurks.
Snakes and rodents may discover the undisguised entrance
to the five-foot hole which the
owls have dug—and follow them
to their lined nest at the end.
The kingfisher's nest (opposite,*

*top) is much safer because
the tunnel may be longer—up to
15 feet—and it is dug into a
steep riverbank. The male and
female may work two weeks to
excavate it. Pecking into the
bank with their beaks, they
scrape the loosened soil back
toward the entrance with their
feet, and sweep it out with
their strong tails.*

The pileated woodpecker (below) uses its stiff tail as a brace while drilling a high-rise hole nest in a Florida cabbage palm tree. Both male and female use their chisel-edged beaks like jackhammers to excavate a cavity large enough to hold both of the 18-inch birds. But even at 45 feet above the ground, the nest may be raided by raccoons and black rat snakes.

Even an eggshell—itself one of nature's most perfect designs—can tell you something of the nest that cradles it. At the end of its burrow, dug several feet into a riverbank, the kingfisher's egg rests secure from hungry eyes. And so it is white, as scientists believe all birds' eggs once were. For birds are heirs to the reptiles whose young still emerge from leathery eggs of white. But as the family of birds flapped and soared into their newfound niches in endless variety, they became specialists in nest designs, most of which leave the precious eggs exposed when both parents must be absent. In such a nest you can read nature's life insurance policy, spelled out in the varied colors, dots, and splotches that make an egg harder to see in an uncovered nest.

What can you learn from the shape of an egg? The kingfisher's is nearly round—and it can afford to be, for at the end of a deep burrow there's no place it can roll. But the murre's egg is sharply tapered at one end; start it rolling and it simply pivots about in a tight circle. That's the shape of survival for an egg laid not in a nest but on a bare rock ledge with a plummet to the sea only a short roll away.

How much safer to nest in a hole. Nature provides some cavities in riverbank and rocky crevice, and more than a few takers gladly move right in, adding only grasses, sticks, or feathers. Others who prefer to nest in tree trunks are clever enough to let another bird drill the hole for them. For this jolting job, the woodpecker is marvelously well constructed. Its stout beak splits and punctures the wood with amazing velocity as the sharp claws on its feet grip the tree's rough bark. Motion pictures slowing the action suggest that it is because the head drives straight back and forth with no sidewise motion that the bird suffers neither brain damage nor whiplash injury. With such formidable equipment, the woodpeckers could easily command first choice of vacant holes. Yet each spring nature's rafters ring as the woodpeckers carve out new ones—and each spring the forest bustles with squatters refurnishing the old woodpecker holes.

Other hole-nesters aren't even fussy about what the hole is in. One family of wrens set up shop in a scarecrow's pocket, another in a sack

Ground-nesting birds devise ways to evade ground predators.

A Wilson's phalarope uses grass (below) and flamingos sculpt with mud (opposite), but both colonial species locate their nests where nestbuilding probably began: on the ground. Natural selection favors ground nesters who hide the nest, increase its height, or build close to others. So the phalarope in a Utah rookery bends surrounding marsh grasses to form a canopy over the eggs. And in Florida, captive flamingos lay their eggs as wild ones do, on earthen thrones 5 to 18 inches high. The nests are built when the ground is wet, each sitting bird stretching its long neck to pull in the mud with its inverted beak.

of nails, still another in a basket of dynamite. Audubon painted a family portrait of wrens at home in a battered hat.

A nest off the ground—whether on a branch, in a mailbox, or atop a traffic light—spells greater safety for its owners. Safety from washouts in the rains of spring, from hungry animals patrolling the ground by day and night, from the unlucky footfall of a passing grazer, from the occasional ground fire that can creep through the duff of the forest floor without harming the greenery above. Yet despite the advantages of castles in the air, many birds still nest on the ground as did their reptilian forebears. And with ingenious adaptations they manage to even the odds against them.

Down here on the ground scurry many of the hatchlings that can flee the nest almost from the moment of hatching. The killdeer starts life as a mottled brownish egg in a crude scrape of a nest among the pebbles or field grasses. Its only defenses are its camouflage and a parent willing to distract an intruder's attention and lead or drive the danger away. If all goes well, out of each egg pops a wide-eyed little scale model of its parent, a far cry from the blind, naked birds born into treetop nests which they can't leave for weeks to come.

The quail of the meadow and the duck of the marsh also tumble from the egg neither helpless nor yet fully self-sufficient. They can run about and find food, but have no defenses against predators. Until they can fly, they can only seek safety under mother bird's shielding wing. But she cannot protect them all of the time, so both species compensate for the inevitable losses by filling capacious nests with eight eggs, ten eggs, a dozen, or more.

Seabirds such as gulls and terns safeguard the species not with large broods but by rearing small broods in large concentrations of nests, sited on windswept rock slab, sandspit, or dune. There is safety in the large numbers of birds in these shore rookeries, especially for the young birds in the center, but scant protection for the eggs in the minimal nests these species provide. A few sticks may be strewn in a shallow depression; a handful of shells and pebbles may be scraped together in the sandy soil to enhance the egg's natural camouflage.

Much greater protection from ground predators rewards the ledge nesters who take to the heights on sea-facing cliffs. Gannets, common murres, kittiwakes, and razor-billed auks crowd onto every level square foot of rock, each pair returning to its own space each summer. The razorbills and murres build no nest at all, entrusting first their single egg and later their naked chick to the windswept bare rock. Gannets at least line with moss a shallow bowl of weeds which they may have stolen from their neighbors, and which both male and female fussily rearrange as they take turns incubating. The smaller, gull-like kittiwake, nesting on the smallest ledges, takes fewer chances; it cements its mossy mud cup to the rock.

Playing it safer still are the crevice nesters who find a haven among the rocks closer to the water's edge. Because black guillemots can thus hide their nests from the omnipresent marauding gulls, the guillemot chicks remain with their parents for six weeks, three times as long as their cousins on the heights.

A surprising nesting arrangement for a seabird is the burrowing of the common puffin. Digging three or four feet back into the slope with his broad, stout beak and shovelling the sand and dirt behind him with his webbed feet, the male curves and slants the burrow downward to about a foot below the entrance. The chicks spend the first six weeks of their lives in this blackness.

There is endless variety, and frequent oddity, in what birds build and how they use it. A hummingbird crafts its little thimble and uses it only once; a pair of bald eagles adds to the same nest year after year until the huge agglomeration weighs upwards of two tons. Such a mass can crash to earth under its own weight —but so can the cup of mud that the cave swallow plasters against a wall and then adds onto without knowing when enough is enough.

Birds may be nature's most innovative builders, but almost all of their nests are used only a few weeks and for one purpose: as a nursery. The adult bird, being superbly insulated and supremely mobile, has little need of permanent shelter for itself. Faced with bad weather, it simply improvises. Grouse in the northern forests sometimes plunge into a snowbank and let

he sifting flakes cover them over, for snow is 0 percent air and thus a good insulator. But ext day the birds explode from their overnight efuge to face the elements unaided once again.

A home in the earth attracts many creatures ecause it offers more than safety from ene-ies above. It affords escape from heat and old, shelter if it's raining, moisture in the esert. In nearly every environment burrowers iddle the topsoil with worm holes, spider bur-ows, bank swallow nests, and other digs.

Many mammals choose a home in the ground -and that seems a little odd. Why, with all eir brainpower and their dexterity, have the ammals shown so little sophistication as uilders? With the single shining exception of e beaver, many seem content to dig a simple ole or go without shelter except as circum-

stance provides. Whatever the reasons, only a few mammals seem able to satisfy their needs in one place for very long. Many who can are rodents—the woodchuck digging into the pasture, the muskrat into the riverbank, the kangaroo rat into the sand.

The prairie dog is another, a burrower nonpareil and a sociable rodent whose "towns" once stretched unbroken for hundreds of miles. Once known as the "plowman of the plains," it kept the soil mixed, aerated, and fertile by bringing up dirt from as far as 30 feet below the surface.

The mole rat of Africa is another determined digger. Its skill has cost it much of its gift of sight, but of what use are eyes in the lightless world of grazers who live by nibbling roots from the ceilings? In this subsurface world it's

Seabirds advance by seniority to choice summer homesites.

Each spring thousands of cliff-breeding gannets return from wintering at sea to Cape St. Marys, Newfoundland (opposite). Older birds force a new pair to build their crude nest of mud, kelp, or lockweed on the edge of the colony, where gulls may steal their egg or chick. Each year they move a little closer to the safer center as the deaths of older birds open up new sites.

Not all ledge-nesters are seabirds. The rare black stork (above) breeds on rocky cliffs in the mountains of Africa.

*Den dwellers find or dig homes
in earth, rocks, or trees.*

---

*A cougar kitten in a rocky den
(opposite) and a pair of raccoon
cubs in a hollow log (bottom)
peer from the rough, unlined
lairs in which their mothers
took shelter to give birth.*

   *Far more care goes into the
building of the underground
burrow of the eastern chipmunk
(below). No telltale pile of loose*

*dirt marks any of its entrances
for it is carried away as fast
as it is dug. In the most elaborate systems, the young are
born in a grass-lined nursery
off the main shaft, between a
well-stocked pantry and a
master bedroom at the end of
a 30-foot burrow. And the
"chippies" are trained to use
the toilet at the farthest,
lowest level of the family's
year-round, all-purpose home.*

the teeth that count, and with incisors like hoes
the mole rat gnaws out corridors, bedrooms,
and food cellars. Cheek flaps close behind the
big incisors to keep dirt out of its mouth, much
as flaps in the beaver's mouth keep out water
when it gnaws branches at the bottom of its
pond. As the lead mole rat gouges the dirt
away, it simply kicks it back to other rats in a
bucketless brigade that ends at the surface in
a spouting cone.

   The mole rat honeycombs the soil and sometimes devours the farmer's crop of tubers, but
for altering the landscape for man and beast, no
diggers—indeed no other wild creatures—have
equaled the impact of the beavers. Their de-
termination to live in a fortress surrounded by
water results in the felling of thousands of trees
and the flooding of uncounted acres of wood-
land and meadow.

   Beavers also dig homes into riverbanks, with
underwater tunnel exits, but they are most ad-
mired for the engineering and lumbering prowess that graces the wilderness with their
picturesque ponds and lodges. To create a pond
of their own, a family of beavers locates a small
creek with trees growing on the shore, pref-
erably their bark-diet favorites—aspen, alder,
willow or birch. Working on the upstream side
of their chosen dam site, they lay a foundation
of stones, sticks, sod, grass, and leaves across
the stream, chinking it with mud they dredge
with their forepaws and carry under their chins.
When the dam begins to hold, the logging con-
tingent starts delivery of slender tree trunks.

   At the logging site as at the dam, no one
seems to be in charge, each beaver is proficient
at all tasks, and all toil harmoniously with no
audible signals. Each tree is felled by the sharp
incisors of one beaver chiselling out two-inch
chips of wood at a bite. Others help to strip the
fallen tree of limbs and buck it into three- to
six-foot lengths. If the logs are far from the
stream, the beavers may dig a canal in which
to float the logs to the water, then tow them
downstream to the dam. There the poles are
heaved over the dam, butt end upstream, to
build a supporting jumble on the lower side.

   The lodge is built next in the same solid fashion, the work going on night after night until
the beavers feel safe for all seasons.

Nature's lumberjacks and engineers create their own habitat.

A beaver dam and lodge in the Tetons (opposite) is 20 years old. Inside the lodge (above) beavers swim in and out of the snug, dry room where they eat and sleep during the day. At night they fell trees and tow the branches to restock their underwater larder. (right)

They built the domed dwelling by spearing saplings into the mud, then compacting more logs with mud and rocks until the foundation stood above the water level. To form the family room, the beavers made the walls thinner as they rose, leaving air holes in the dome.

Kits (left) are born in the lodge in the spring. In six weeks they are weaned to a bark diet; by autumn they are helping with the family chores.

The year-round, all-purpose homes of the prairie dog, the mole rat, and the beaver are exceptions in the mammal clan, however. The wolf, the fox, the bobcat, and most of the bigger mammals are opportunists who seldom look for or dig a temporary shelter except when they sense the oncoming birth of their young. Many insects, by comparison, build elaborate abodes not only to serve as nurseries, but for protection from enemies, for shelter from the elements, and for food processing and storage. In the anthill, the wasp nest, the beehive, and the termite mound we see some of the most productive, cooperative organisms on earth. In the well-known works of these social species we humans observe—sometimes with a twinge of envy—great cities in miniature, ticking along more smoothly than our own.

In the microworld of six- and eight-legged builders the solitary workers also turn in stellar performances, especially when they build to catch a meal. As breadwinners, these tiny artisans field an impressive array of traps, nets, and snares. And none can build them better than the spiders.

All spiders spin silk and nearly all have poison fangs. For most of them, the poison subdues the victim *after* it is caught; the initial offensive weapon is the silk. And what wonders they weave with it! The ogre-faced spider spins a snare the size of a thumbnail and casts it over a passing victim. The bolas spider dangles only a single strand of silk, swinging it to-and-fro until a sticky globule on the end tacks onto a passing insect which it then reels in.

Humans seldom see these spider specialists at work, but the familiar orb weavers are all around us. We marvel at their delicate spirals pearled with the dew of a chilly fall morning. And sometimes, we fume at the errant web that appears overnight in the corner of a just-cleaned room. The room may have been pitch-dark, but the web appears nonetheless, for the weaver works not by sight but by touch.

In the early hours of morning the spider begins its task. The silk with which it will build is a liquid protein that stiffens as it is drawn from pairs of nozzles called spinnerets. Protruding from the underside of the abdomen, they shoot out a strand that anchors itself to the

Web and cocoon spinners use materials from their own bodies.

A female orb weaver rests from building the most advanced of all spider webs (opposite). Using a minimum of silk to cover a maximum of space, she probably spun the web in 30 minutes and now waits for it to entangle the first insect. If the web is torn in the struggle, the spider will repair it or build a fresh one.

Another silk-spinner, the luna moth caterpillar, first selects an oak leaf (right) or a sweet gum (bottom) or other leaf to curl about itself, sealing the edges with silk. Inside the leaf-tent, the caterpillar turns its head to direct the flow of silk to form a cocoon. When the caterpillar has become a moth, it secretes a substance that weakens the cocoon, so it can work its way out and fly away.

91

first solid object it touches. The spider then reinforces it by crawling along the strand and paying out a heavier line, the true bridge strand. The spider anchors the other end to its perch, and the bridge strand is ready to support the construction of the rest of the web.

Now the tiny builder delineates the web's perimeter and lays down radials, working from the outer edge to the center. Fixing the hub with a closely woven platform, the spider now spirals back out from the hub, laying a continuous strand in ever-widening circles.

At this point, a fly could blunder into the web and buzz away without missing a wingbeat —or becoming a meal. The web thus far is bone dry; only now is the spider ready to add the deadly sticky spiral that will do the actual catching.

This spiral is a tough cable coated with stickum, stronger than steel wire, size for size. It's only one of six known varieties of spider silk, although no one spider produces all six kinds. To lay it, the spider spirals inward toward the hub, feeling along the previous guide spiral with its front feet, and tacking down the sticky strand with its hind feet. As the spider goes, it neatly rolls up the guide spiral, segment by segment, and eats it or discards it. Like an artist signing his work, the builder finishes with a snowy-white zigzag across the hub. In fact, it almost *is* a signature, for its design alone is sometimes enough to identify the species of the maker.

If another spider enters upon the finished web, the owner may claim the interloper as a meal as readily as if it were a grasshopper or a butterfly. For nearly all spiders are predators, loners, and cannibals. Even the act of mating puts the diminutive male in great peril of his life from the much larger female.

Beneath the ocean waves in the warm water lagoons of the world we find age-old building projects that rival the delicate intricacies of the spider's web and surpass the size of the beaver's dam and lodge. From the tiny skeletons of coral colonies grow vast reef communities of incredible beauty and fantastic variety. Each individual living polyp is no bigger than the point of a felt-tip pen and it uses invisible tentacles to capture the plankton it feeds upon.

Eventually it leaves its tiny deposit of calcium carbonate on one of thousands of branches or stalks that make up the ever-expanding reef.

The coral reef itself is home to other individual builders who must cope alone with all the dangers of undersea life. In the reefs off Australia the female gall crab rasps out a chamber somewhat like that of the woodpecker: large cavity, small opening. She takes up residence inside, and there she feeds and fattens by straining food particles from the water. The day comes when she cannot get out the door. In her cell of coral she serves out her life sentence, beautifully protected from potential enemies, yet accessible to the smaller male when he drops in to mate.

Where sea meets shore the lugworm builds to hide—then gives itself away. Its burrow is a U-shaped affair with the worm at the bottom and an opening at each end. The worm sucks food-laden water in one end, then expels its casts in a neat little mound at the other end— a sure tip-off to worm eater and fisherman alike that there's a lugworm to be dug up here.

On the same beach you may find countless numbers of a different kind of marine builder— those whose architecture is lavished upon their own bodies. Almost any wave will wash ashore a testimonial to the artistry of clam and conch, of scallop and nautilus.

Fish take cover under rocks or logs, in seaweed, coral, or the hulks of wrecked ships, always counting on their marvelous mobility to reach these shelters when needed. Few fish need a nest for their young, but for those that do, a simple hollow in the sand or mud bottom is usually enough. But not for the three-spined stickleback, a small but admirable fish that outshines all others of its kind as a builder of nests. In a shallow scrape on the bottom of a stream or pool, the male stickleback assembles an assortment of weed snippets. Using a gummy exudate from his own kidneys, he glues the mass together. Next he burrows into the mass, using his body as a ram to punch a tunnel through. That completes the nest; all that remains is to lure a procession of females into it, fertilize the eggs they lay inside, and then spend the next week or so fanning fresh water through the tunnel to keep the eggs aerated.

Work of underwater builders resembles land architects'.

Few fish build nests, but two that do use surprisingly bird-like methods. A female sockeye salmon (opposite, top) uses her tail to scrape out a crude ground nest. In it the male standing guard (above) deposits milt and the female lays her eggs, covering them with sand. The male three-spined stickleback (opposite, below), nudging his mate to lay eggs, built their nest of aquatic plants. Like a chimney swift, he bound the plants with a sticky substance from his body.

More like a snail shell is the home of the chambered nautilus (opposite, center). The nautilus enlarges its home by secreting calcium carbonate in semi-liquid form from the upper layer of cells in its soft body. Added to the open edge of the spiral-shaped shell, the calcium carbonate hardens quickly in salt water. The nautilus moves forward, leaving a space which is then sealed off with a curved partition. A small hole in each partition allows a tube of flesh to vary the volume of nitrogen in the empty rooms, controlling vertical navigation.

From ocean bed to mountain crag, the animals' search for shelter goes on. No inviting opening is overlooked. A few creatures who prize extra warmth and shadowy concealment find all their site requirements under man's own rooftree—and have the courage to move in. Swifts line old chimneys with semicircular stick nests which they glue to the brick with their own saliva. Barn swallows attach their mud cups to the rafters, wasps hang their paper castles under the eaves, and hungry spiders hide their snares upstairs and down.

With so many builders building so many kinds of structures in the varied environments of the planet, it should not surprise us to find a few coincidences now and then. Yet we exclaim when we find an alligator in Florida building a nest mound that will be warmed by compost, almost exactly as does the mallee fowl in Australia half a world away. We know that a snarl of snakes may share an underground nest in winter, but we are surprised to learn that scores of South American finches do the same in the Bolivian Andes.

Surprising contrasts also abound. Some of the penduline titmice build nest after nest, enticing females to raise a family in each one, while on bare ledges some seabirds build none. And in the frigid Antarctic, the emperor penguin copes with the scarcity of nest-building materials in quite a different way—the male incubates the pair's one egg by holding it on top of his webbed feet, keeping it warm for weeks in the fold of his underbelly.

There are also contrasts in the way different creatures avoid the chore of homebuilding even where nest-building materials abound. The cuckoo simply sneaks its egg into another bird's nest—a neat parasitic variation on the familiar squatter's strategy, so ably employed by the hermit crab, the field mouse, and the many birds that nest in old woodpecker holes.

Most squatters are at least prudent enough to move in only after the original builder has vacated the premises, but not the rufous woodpecker of India and Ceylon. In a nice reverse twist of its usual role, this little woodpecker moves into the football-sized nest of the black tree ant. The woodpecker hollows out one part and lives inside while the ants are still in

A few wild creatures prefer to build close to man's shelters.

Living wild and free amid the comforts of domesticity, the white stork (opposite), the barn swallow (below), and the common wasp (bottom) appear to have the best of both worlds. Whether it was warmth or hiding places that lured each one out of the trees to live with man, the pattern is an ancient one.

The stork's reputation for bringing good luck led Europeans to try to attract more of the long-legged birds by making roof platforms on which they could build their bulky nests of sticks.

Beginning to build where man leaves off does not necessarily lessen the work, however. One pair of barn swallows may make hundreds of trips carrying mud pellets in their trowel-shaped beaks to build one nest.

The wasp's sting makes it a less welcome tenant, but man is forced to admire its ability to make a rainproof, paper home out of chewed old wood and tough plant fibers. Mixed with the wasp's saliva, the fibrous paste is matted and dried into paper, essentially the same process used in paper mills.

Squatters capitalize on efforts of other wild builders.

The white-footed mouse peering from an abandoned woodpecker's nest (below) knows a housing bonanza when it sees one. Although the little squatter lined the hole with a cache of cattail seeds, that was far less work than building a roofed grass nest on the ground.

To build or not to build is not a question for land hermit crabs (opposite). Their ancestors chose to recycle the cast-off shells of sea snails so long ago that the crab's abdomen has become twisted to fit the spiral shape of the snails' shells.

The little beachcomber shown here is wearing a Florida crown conch shell, which it will

discard for another when outgrown. It won't contest a tenant for its shell, but two crabs will fight over an empty one.

The young cuckoo (bottom) being fed by a reed warbler in whose nest the mother cuckoo laid her egg, is more a foundling than a squatter. The warbler works hard to feed the oversized "guest"—as if unaware that it has evicted her babes.

residence. The ants are fierce defenders of hearth and home, yet for reasons still unknown, they accept the havoc and occupation without protest. And the birds, for reasons equally puzzling, allow the ants to come and go without molestation—even though the black tree ant is the staple of their diet!

As we overbuilding humans gradually elbow our wild counterparts aside, we can perhaps take some comfort in yet another surprising fact: we have not driven them all away from their forests and their meadows. Indeed, some of them have adapted to our alteration of their domains rather well. Witness the growing numbers of robin nests in our yards as their builders flock to the ever-expanding worm-hunting grounds on the lawns of suburbia. Witness the pigeon moving its sorry nest of random twigs from our torn-down barns to the ledges and lofts of our urban canyons. Witness too, the many wild builders that find in our litter the stuff of which contemporary urban nests are often made: string, wire, rags, paper, even a soup can with the right size opening. One osprey nest featured an arrow, some pilfered laundry, and a toy garden rake.

Long before there were humans to adapt to, there were animals seeking shelter and either constructing it or finding it in the workmanship of other animals. Today that search and the endless adaptations go on. That same osprey nest, for example, may well become an apartment house, sheltering in its lower levels a family of sparrows or wrens. Squatters are often tolerated by both ospreys and eagles, even when the tenants forage in the penthouse while the owners are away on a hunting trip.

From nature's building projects, human architecture has probably absorbed more than we realize. Certainly we could profit by taking a closer look at their guiding principles: form follows function, little is done without a reason. What animals do by instinct, man has painstakingly learned to imitate, adding only variation and ornament. As our appreciation of energy grows, we may yet learn to emulate nature's economy. Building projects—both wild and human—evolve to fit new necessities. Adaptation is the key to survival and nowhere is this more evident than among the builders.

Finding a Workable Lifestyle

The gregarious ones reap nature's social security benefits; other wild creatures face life alone.

The Cooperators

Ton upon lethargic ton, the huge sea animals sprawl over each other. The leathery mass twitches, writhes, bristles with stout tusks. The walrus bulls are trying to sleep—but how can they in that jampacked mob? And how could they flee if danger loomed? Sharing their body heat helps them to survive on their icy islet. Yet for centuries their habit of huddling in great numbers has made them vulnerable to their chief predators, men who saw in their massive bodies oil, ivory, and hides. By crowding together the walruses lose some advantages but gain others in exchange. That is the nature of cooperation, and by its ambivalent terms a wide range of wild creatures must abide every day of their lives.

But there are many ways to cooperate. By working together, a pack of wolves can drag down a full-grown moose, a feat a lone wolf will seldom even attempt. But each wolf is also capable of taking independent action, so the pack at its best is often wrenched with squabbles and rivalries. Compared to the clockwork of an anthill, the inner workings of the wolf pack seem a fitful contraption indeed.

To find one ant is to find hundreds, even thousands, for these creatures in all their species are so committed to togetherness that none can go it alone. From their bustling societies scientists have drawn a wealth of insights into social behavior, the life-style of the gregarious cooperator.

But if we would find nature's first and most primitive cooperators, we must look beyond the anthill to the life that is visible only through a microscope. Prowling the soil beneath our feet —or the glass slide beneath our lens—the shapeless amoeba flows over its microscopic prey. About every four hours it divides, and two amoebas dine on bacteria where there had been one. If unchecked, 20,000 explode to more than a million in only 24 hours.

Loners all—until the food runs out. Then an amazing process begins. As if to save themselves from starvation, the microscopic creatures actually fuse their separate bodies together to form a new animal, a single multi-celled creature called a slime mold. Because the new animal is more mobile than a single amoeba, it is more likely to reach a new food source. Eventually the mold releases spores that germinate into new free-living amoebas. Together these smallest living animals do what none could do alone. And that is usually why animals cooperate.

Amoeba, ant, wolf—each shows us a different kind and degree of social organization. The amoeba surrenders its very self to the mass. The ant remains an individual, but with a specialized body structure that fits it for only one role and ensures its dependence on the group. In fact, the group, not the ant, is the complete organism. Wolves are highly developed individuals capable of performing—and even competing for—more than one role. Yet each way is good, for it serves its practitioners' needs in the struggle to survive.

In the hive of the honeybee we find the same division of labor among cooperating specialists —the same variation of form within a single species—as we find in the anthill. Every hive has an outsized queen, up to a few hundred males to fertilize her, and some 50,000 sterile females to do the work. But among the female bees it is age as well as anatomy that allot the needed chores.

Fresh from her larval cell, the new worker bee pitches right in, carrying wastes from the hive, cleaning out empty cells, even sanitizing them with a disinfectant that is secreted by her body. In three to six days she becomes a baby-sitter, feeding the larvae with honey and pollen and a special substance her body now provides. At about 12 days she's a builder, her body now a wax factory. At 20 days she guards the entrance, and soon leaves it to forage until she dies, at the age of four to six weeks.

Impressive though the bees' accomplishments are, the social organization of the ants is as complex and their output more versatile. Also, they live for years while the bee's life-span is measured in weeks. Many species of ants work within the same basic caste system of queen, workers, and males, to cultivate fungus "farms," keep aphids as "cows," capture "slaves," and maintain "armies."

When the mated young ant queen has found a nest site and hatched her first generations of workers, she soon has a full "court" flourish-

Teamwork enables wolves to bring down big game animals.

Tense and alert, a wolf pack silently closes in on a moose. The cow may react by standing her ground, advancing toward the wolves, or trying to escape at a fast trot. Wolves usually hesitate in a face-to-face challenge, but almost invariably chase frightened, bolting game. Long leaps and bounds help them to outrun their quarry.

The pack strikes first with powerful bites high on the rump to avoid the cow's hard, sharp, hind hooves. Digging in with two-inch fangs, they hold on as long as they can while she plunges on. One wolf may grab for the bulbous nose, giving others a chance to tear into the shoulder and neck—staying clear of the front hooves. A healthy moose may shake off the whole pack only to succumb to renewed attacks. Continuous harassment finally causes the exhausted victim to fall and death soon follows.

Previous page: Walrus Bulls—Round Island, Alaska.

Efficient insect families share
the chores of food gathering.

Army ants, leaf-cutter ants,
and honeybees cooperate in all
family tasks. Worker army ants
(above) build a bridge of their
bodies to reach a wasps' nest.
Three "soldiers"—those with
larger, lighter-colored heads—
guard the attacking army. Using
the bridge offers advantages
over crawling across the leaf:
it gives the ants firmer footing
and allows them to travel
right side up, making it easier
to carry prey. Raid columns
cross the bridge and rip holes
in the paper nest to capture
pupae for the family larder.

Unlike the army ant
"hunters," the leaf-cutter ants

(top) grow food by "farming."
The larger adults use sharp
jaws to cut wedges from petals
and leaves, while smaller
worker-ants protect them from
attacking flies. Other workers
waiting at the nest lick and
chew the leaves to hasten decay
before spreading them on
compost. Another group of small
ants remove unwanted spores
from the fungus crop.

Honeybees neither hunt nor
farm. Instead mature workers
forage afield for nectar and
pollen, while younger workers
(opposite) stay in the hive to
build wax cells for housing
larvae and storing honey.

ing around her. Soldiers and workers scurry
about their appointed tasks. The worker ant
gathers the food, feeds the larvae, tends the
queen, cleans the nest. The soldier, huge-
headed and big-jawed, stands guard over the
workers and the nest. Intruders soon learn it
will sink its stout pincers into anything that
threatens. But the soldier cannot forage for
itself, so that chore is added to the worker's
job description. And, with rare exceptions,
neither worker nor soldier lays eggs. This high-
est of all responsibilities rests with the queen.

Intriguing adaptations vary the basic plan.
Some guard ants have heads that are broad-
ened, flattened, and even colored differently
to serve as inconspicuous plugs in the entry to
the nest. Some workers serve as living storage
jars, hanging from the ceiling with their swol-
len abdomens full of nectar, awaiting some
future time of need. And some larvae even find
service as needle and thread; adults of the tailor
ants pass the larvae from leaf-edge to leaf-edge
while the larvae spin silk to sew the leaves into
a nest. So proficient are they that this unique
shelter will last through weeks of wind and
weather.

But nothing we observe in the efficiency of
the underground ant nest or the tree-dwelling
ants' leafy retreat equals the accomplishments
of the terrestrial army ants of tropical America.
Of no fixed address, they live their entire lives
on a forced march of thousands across the forest floor. Big soldiers with jaws like ice tongs
march on the edges of the column. Smaller
workers trudge with their gigantic queen; some
lug helpless larvae.

At a suitable site, perhaps a hundred yards
from the morning's starting point, the workers
create a night's shelter for the colony. First
they anchor themselves to a chosen branch or
log and then to each other until they hang in
chains, clusters, and solid masses; eventually
they become a seething tangle measuring up to
three feet across.

Nature supplies its many architects with all
sorts of building materials, but none erects a
home more bizarre than this one. In fact the
workers *are* the nest; it contains no building
material but their own living bodies. Some of
the ants support hundreds of times their own

weight. Yet they seek no relief until the hour is right for moving on.

That may be the next morning or three weeks later. Periodically a bivouac is extended while the queen settles down and lays several hundred thousand eggs. About 20 days later a new generation of workers will have been hatched and reared and made ready to take their places in the next nomadic episode.

Meanwhile some of the workers, guarded by soldiers, spend the days out raiding the surrounding tropic forest floor for food. Wasps, other kinds of ants, and other social insects, an occasional roach, scorpion, and tarantula—all are fair game to the carnivorous army ant on the prowl. The first worker to make contact clamps on to the victim with a paralyzing sting and signals for reinforcements with a chemical alarm. The signal is relayed through the ranks; more workers surge forward to the skirmish. Together they kill the victim and carry it back to camp. If too large to carry, it is torn to shreds first. Other workers resume the raid.

It is always by dark of night that these marauders of the forest dismantle their living nest. In minutes the yard-wide bivouac melts into a river of ants. By midnight, the workers have again anchored themselves, first to a deadfall and then to each other, until they have formed a new shelter in which the queen may rest through the next day. At night they move on again in a series of one-night marches that ensure fresh hunting grounds each day, until it is time for the queen to produce the next generation. And amazingly enough, the entire operation is regulated by chemicals, for the terrestrial army ant is so nearly blind it can probably only distinguish light from dark.

This is more than just cooperation, more even than a well-organized society. Like the bees in their hive, the ants are in reality a family, a mother and her quarter-million or more children.

Man has always been deeply impressed by the degree of social order he observed in the earlier, simpler forms of life. The explanation lies in their close kinships. Amoebas multiply by dividing; the two are literally of one flesh. Ants and honeybees are of one family; each is sibling to every other in the group.

Cooperation in homebuilding is tailored to animals' life-style.

Nomad army ants (opposite) and stay-at-home prairie dogs (below) devise shelters to fit their totally different needs. The ants bivouac briefly in a temporary living structure formed from their own bodies.

To construct a quick home, worker ants hanging from a branch or fallen log hook together the claws of their hind legs. Within 4 to 6 hours a mass of 100,000 ants forms a frothy, lacework curtain. The spaces formed by the ants' linked bodies become perfect chambers for the queen and her soon-to-be-born brood.

Prairie dog homes consist of a complex network of under-ground burrows with nesting chambers which everyone helps to build, expand, or repair. With the excavated earth the adults form mounds near the entrances to keep out flash floods. The mounds also serve as sentry posts for the rotating family guards on the outlook for trespassers.

In sharp contrast the vertebrates, with their wider mating contacts and mixing of genes, tend more toward variation in their offspring. The earlier social orders run so smoothly because cooperation is automatic, pre-programmed into the species. Among vertebrates cooperation depends at least partly on learning.

When the prairie dog pup first emerges from its natal burrow, it scans a world of grassland pocked with burrows in all directions. But it soon learns that it dwells within an invisible fence. A few neighboring burrows belong to members of its "coterie," and these are within its fence. But if it ventures beyond, it will be warily approached and tested with a kind of kiss to determine whether it belongs where it has wandered. If it flunks this identification check, it is chased home unceremoniously.

In centuries past, prairie dog towns tunneled the Great Plains from Canada to Mexico. Even with all that space, the towns were little more than vast accumulations of little coteries, each headed by a male that had won it by a show of strength, and each passing its boundaries along from generation to generation.

In this most elaborate of rodent societies, ritual helps to bond the coterie together. When a kiss identifies a friend, both prairie dogs often will sprawl out and begin to groom each other. In so doing they remind us of the monkeys whose urge to groom is a hallmark of that much-studied social group. Though half a world away, the advantages are the same. Grooming helps keep the group clean and healthy, and it eases the tensions that are part of the price of togetherness.

There is safety in numbers, and animals beyond numbering seem to know it. The desire to thwart predators and to avoid facing one alone can roil a stretch of ocean with a sprawling school of fish or set a plain to trembling under many kinds of hooves. But school and herd are often only accidental clumpings of animals who happen to be fleeing the same foe, going to the same place, or exploiting the same resource. They lack the leadership and organization of more permanent animal societies, yet they attain together benefits they could not enjoy by traveling alone.

Such chance grazing groups find the giraffe to be a valuable companion because of its lofty viewpoint and wide range of keen vision. Around its stilty legs the zebra, impala, hartebeest and wildebeest frequently gather. When browsing at treetop level, this living lookout tower regularly lifts its head to survey the plain between bites. If something startles the giraffe, fellow grazers and browsers take off with it without waiting to see what startled it. Together the hurtling horde can befuddle an attacking trio of lions with a wild confusion of shapes and sizes and gaits and bellows and scents.

The gentle giraffe seems to give more security than it gets from its short-legged companions—except when they meet at the waterhole. All animals approach the waterhole warily, knowing that predators often lurk nearby, but the giraffe is especially vulnerable. It must spread its forelegs far apart as the head goes down, rendering it momentarily helpless. Yet it may be warned in time by the alarm calls of the others, a favor inadvertently returned to their tall benefactor.

While many predators prefer a one-to-one encounter with their prey, others show no interest in prey *except* when they are present in large numbers. The tuna and the striped bass attack only schools of fish. Their victims are unfortunate exceptions to the cooperators' rule. Nevertheless, on balance, the principle holds true: whether schooling in the sea or herding on the land, traveling with a crowd provides a moving buffer zone of flesh for most of the individuals and increases their margin of safety from predators.

Fleeing together from danger increases animals' security.

A school of black-striped grunts (left) stays close together for protection against predators. When one senses danger and darts away, others pick up its vibrations and switch course with it. One out of four species of fish seek safety in society for at least part of their lives. Some schools are made up of a female with a multitude of her newly-hatched fingerlings that remain near her for protection.

When alarmed, giraffes too take off together at top speed (below). The whole herd can be mobilized by a single panicky sound or movement: a snort, a perpendicular head, flared nostrils, blazing eyes, stiffened ears, nervous pawing, or tail switching. When escaping with young, giraffes decelerate from 40 to 25 miles per hour to accommodate them. The adult's size, speed, and kicking power give it immunity from most predators and save many of its calves from lions, the leading cause of their death.

Social animals live together in degrees of organization that vary with their needs; among the most interesting are those that have a leader. When watching a school of fish, you will search in vain for one. There's the leader now, you may think, out front at the head of the class. But watch as the school changes direction. The fish all pivot as one; the lead shifts instantly to whoever is now out front, and the deposed frontrunner is again just another fish in the crowd. In fact, it never led at all; it simply started first.

A school of fish then is not a society in the sense that a herd of elephants is. For elephants have both organization and strong interrelationships—as well as someone who is acknowledged to be the boss.

An elephant herd is actually an extended family, a tangle of intertwining bloodlines linking uncles, aunts, cousins, siblings, parents, and grandparents. Motherhood and a long childhood bond the females to the young; the need for security cements them both to the larger group. But the bulls drift in and out at will; in effect, they're not home enough to lead. So the herd matriarch or lead cow reigns over the clan as a whole, delegating some of her authority to younger females. They, in turn, appear to accord her both respect and affection.

Except for occasional clan reunions at the waterhole, the herd splits into smaller family groups for the day-to-day business of living. In peaceful times, several family units browse within earshot of each other, rallying quickly around the matriarch at the first sign of trouble. She responds to the intrusion with a fierce threat display and makes the decision for the herd to flee or fight. It is almost always the matriarch who leads the herd in an all-out charge or who stands and faces the foe to cover the retreat of the rest of the herd.

In the wolf pack, it is the dominant, or alpha, male who leads. He feasts first at a kill, sharing this privilege with only a few other members. He exercises dominance in mating, sometimes allowing the alpha female to mate with other high-ranking males. No two members of the pack have the same social standing; females as well as males know their place—and flaunt it. But when the alpha male holds his bushy

The role of the leader is to defend and order family life.

In the elephant clan it is usually the oldest reproductive female—the matriarch—who oversees the herd and makes decisions on matters that are vital to its survival and emotional well-being.

During the day, the herd breaks up to graze. Younger females accompanying half-grown youngsters roam far afield, leaving close-by grasses for nursing mothers and calves. Older bulls travel about freely.

At night, some of the cows stand guard over the sleeping herd. In a crisis adults form a defensive cluster, with young tucked between their legs or hidden behind the bulky bodies. Outside the circle, facing the enemy, the matriarch asserts her leadership of the clan.

Wary, security-conscious musk oxen (above) and Canada geese (opposite) nearly always travel and feed in groups. Each species has also devised additional, strikingly different, defense strategies.

At the first hint of danger, the adult musk oxen circle quickly about their young, enclosing them in a formidable barricade of flesh spiked with sharp, recurved horns. By facing in all directions at once, the oxen leave their foes no safe avenue of attack.

Anticipating danger at migration rest stops, the geese post sentinels so the rest of the flock can feed and sleep. A nasal warning call will trigger an instant liftoff that saves the flock from ground predators. (The banded legs of the sentinels in the photograph indicate that a biologist succeeded in beating the flock's sentry system—perhaps by taking advantage of a previous molting season when the birds could not fly.)

tail erect, all other tails must droop, for that is a sign of submission in the hierarchy of the pack.

Though the alpha dominates, he also delegates. To the beta male, his ranking lieutenant, fall the routine decisions of the day. The beta usually leads the way in travel and breaks up minor spats. His on-the-job training spells security for the pack, since he will likely succeed the alpha if calamity ends the leader's reign. If that does not occur, there is little chance for advancement until the weakness of the aging leader forces him to submit to the beta's growing challenge. Yet any young wolf can become an alpha. He needs only to strike out on his own. But he must be strong enough to establish his own territory, win his own mate, and start his own pack.

Nature offers us few opportunities to observe the strategies of two animal leaders of different species when they meet in the inexorable dilemma of predator and prey. Such a confrontation does occur in the high Arctic where tundra wolf packs stalk herds of musk oxen.

Wolves usually focus on the straggler in a grazing herd, but with musk oxen their best opportunity lies in catching the whole herd off-guard. The moment the wolves' presence is detected, the calves will all but vanish inside a fortress of woolly bodies facing outward. For thousands of years that defensive circle of adults, under the leadership of a big bull, has protected the herd from grizzlies, polar bears, and wolves. Any intruder who ventures too close to the circle risks being gored in a ground-shaking, horn-slashing charge by first the lead bull and then the other defenders in turn, each returning quickly to its place in line.

Although the wolves may persist briefly in their efforts to grab a calf, even the tough alpha wolf soon leads his pack away once that horn-rimmed circle has closed on their hoped-for dinner. The musk oxen most often taken by wolves are probably old males found wandering alone. No longer able to compete for mates, they tend to wander far from the herd.

Tragically for the great herds of musk oxen that once roamed the world's Arctic pastures, the defense that worked so well against the wolf pack played directly into the hands of

men; the cluster of shaggy beasts made perfect targets of themselves, first for natives with bows and arrows and later for the hunting rifles of market hunters. Now protected, these hold-outs from the ice age are slowly backing away from the brink of extinction.

Man also forces migrating waterfowl to pay a price for sociability. Migrating flocks of Canada geese scan the scenery below for likely spots to splash down and feed. Spotting their species' distinctive color pattern sprinkled on a reedy marsh, they wheel in for landing—and discover too late what decoys are for.

Those geese that make it down safely find other predators awaiting them on the ground. Because of their large aggregations, the noisy, highly visible flocks may attract foxes, skunks, raccoons, and other four-footed hunters. To help offset that risk, the Canadas post sentries —as do many other gregarious species—to keep a sharp lookout while the rest feed or nap. At a warning signal from a sentry, every head is raised. If the threat warrants, the flock lifts off with a great flapping and cackling. So dependent are Canadas upon the reassuring presence of a sentinel that a single goose, forced to feed alone, may be too busy glancing about to eat its fill.

Another price that must be paid for the rewards of togetherness is an increase in internal strife, especially among mammals. When an animal must constantly defer to another in eating, drinking, mating, even sleeping, sooner or later patience wears thin and the hackles rise. If unchecked, the conflict could waste enormous amounts of energy and cause injury— even death. So substitute rituals have been developed to dissipate the hostility in harmless ways. Even the head-butting contests for territory and harem in the less-structured life of herd animals usually last only until it becomes clear who would win if the tussle turned ugly. Once that has been determined, what need is there to battle any further? The weaker surrenders—but he lives.

Sometimes the ritual gives way to real fighting, but both combatants usually fight fair, and so both usually survive. Thus the herd benefits from nature's honor system, whereby even in combat there is cooperation.

Nurturing the young and nursing the old strengthens family bonds.

When one pelican takes flight, most of the flock is sure to follow (left), often maneuvering their big bodies into formation as they flap and glide away on a fishing trip. Their custom of leaving "babysitters" with the chicks and the patience with which the returned pelicans allow the young birds to probe their gullets for the day's catch have earned the sociable birds the reputation of being good parents.

Strong family ties of a very different sort are evident in the way elephants (above) express their concern over a sick or dying cow. Attempting to help her, family members will put their trunks in her mouth, nudge her, try to lift her, feed her grass or even mount her. When death comes, the elephants mill around for hours to defend the carcass. Other elephants may join them as they return again and again out of grief or curiosity, finally performing the strange ritual of carrying off the tusks.

Like a blizzard running backwards, great snow-white birds swirl up from a crowded rookery, filling the sky with chunky bodies gracefully borne on nine-foot spans of wing. For a moment the pelicans form a living canopy over their Canadian island nursery before they depart on a feeding flight over the surrounding lake. They need not worry about the young birds on the ground, for in this parental co-op adults take turns serving as sentinels and babysitters. Those on the flight also team up to forage for the fish they must bring back to fill all the hungry young mouths.

Unlike its cousin the brown pelican, the white does not dive for fish. Instead, teams of four or five alight on the water in formation and swim shoreward side by side in a crescent-shaped line. In the shallows, they slap at the water with their broad, powerful wings and kick mightily with their feet to drive the fish toward the shore. Closing the arc at each end, the huge birds snap up the panicky fish as they try to zip past them to safety. Back to the rookery the big birds flap and glide, only to be mobbed by hungry chicks eager to ransack the first fish-filled gullet they come to.

The family life of pelicans reminds us of the enormous amount of energy that nature lavishes upon the young because they are the species' investment in tomorrow. In most animal groups, the old are treated as the leftovers of yesterday. They have lived long enough to reproduce and must thereafter—like the sick, crippled, or wounded—fend for themselves as best they can. But in a few societies such as the elephant herd, the old are held in esteem. In the fullness of their years a wealth of wisdom resides, and by their value to the group their right to lead is acknowledged to the end. And so the herd takes care of them, slows to their pace if need be, and comes to their aid in sickness even if risk is incurred. More than once, dozens of elephants have been seen to cluster around an older comrade in distress, flinging dust, trumpeting, and thrashing bushes in a tantrum of threats until a skulking lioness or a photographer's jeep retreated.

Loners as well as cooperators provide somehow for the young. Only the cooperators have the capacity to care for the old and the afflicted.

Here and there in nature's list of cooperators a mismatched set turns up, a strange partnership between two different species that somehow works out fine. Among the cooperators, these are the odd couples. It may begin when an impala finds a tiny pest lodged in its skin. The two creatures are living together, but only one benefits from the arrangement; the other is harmed. So the relationship is parasitic—until the impala teams up with a bird that will keep its skin free of such freeloaders. The impala is relieved of its discomfort and the bird, an oxpecker, gains a steady food supply.

Another unwilling host, a large, bottom-dwelling saltwater fish called a grouper, trades similar favors with a pair of neon gobies, tiny gold fish smaller than the grouper's eye.

In these and other alliances mentioned earlier in which both partners benefit and neither is harmed, the arrangement is called mutualism. When the two mutualists live together, it is called symbiosis. Hermit crab and bristle worm are sometimes roommates in the same shell; the worm cleans house and gets paid with the crab's table scraps. The little melia crab kidnaps its partners; when it is young it grabs a sea anemone in each claw and hangs on for dear life—literally. It never lets go. Yet the anemones benefit all their lives, since the crab carries these normally stationary creatures to food sources they'd never tap on their own. In return the anemones' stinging tentacles protect and camouflage the crab from its predators. Sometimes it is the anemone that hijacks the crab, pasting itself to the crab's upper shell to ride its host piggyback through life.

Before these strange partnerships are formed, it's important—especially to the tiny gobies and other cleaner fishes—that both parties understand each other. By their gaudy hues and ritual antics, most cleaner species make it clear that they are not to be eaten. By ritual movements of its own, the big fish signifies its agreement to the terms, and the bargain is struck. But cheating does occur; now and then a host fish gobbles up its cleaner. And sometimes other small fish mimic a cleaner in color and action—only to take a bite out of the big fish. It's enough to shake a mutualist's faith in any species but its own!

A fair exchange of services creates strange partnerships.

A Nassau grouper (opposite), suffering from parasites and fungi in its skin, has attracted a "cleaning crew"—a tiny Spanish hogfish and a pair of gobies—to eat the invaders and tend the wounds to prevent infection. In return, the little fish are assured a steady food supply and a degree of protection against their predators.

Such mutually beneficial partnerships also thrive between members of totally different groups of animals. In Africa a red-billed oxpecker (left) rids a male impala of its bothersome parasitic pests, cleaning the wounds by removing blood and dead tissue. When danger approaches, the bird flies up and cries a shrill warning to its host. In exchange, the oxpecker receives a dependable diet of ticks and flies and collects sleek impala hair to line its grassy nest.

Solitary predators seek out prey they can handle alone.

A snake, a leopard, and a tarantula reveal the prowess that enables them to eat well although they hunt alone. Each begins with an ambush and ends with a pounce.

The eastern river snake (opposite, top) grabs a green frog, opens wide, and swallows it headfirst, alive and kicking.

The tarantula (bottom) lunges and wraps its legs tightly around the anole while giving it fatal injections of poison followed by digestive enzymes. Within a half-hour the tarantula starts to suck the liquefied tissues from the anole's body.

The speed and fangs of the leopard will make short work of the doomed bush pig below. Hauling its kill into a thicket, the leopard feasts as it hunts—alone. It often must make do with small game—a bird, a fish, a snake, or a monkey. When it brings down larger prey, such as the pig or a gazelle, the leopard hides leftovers in a tree fork for a future meal.

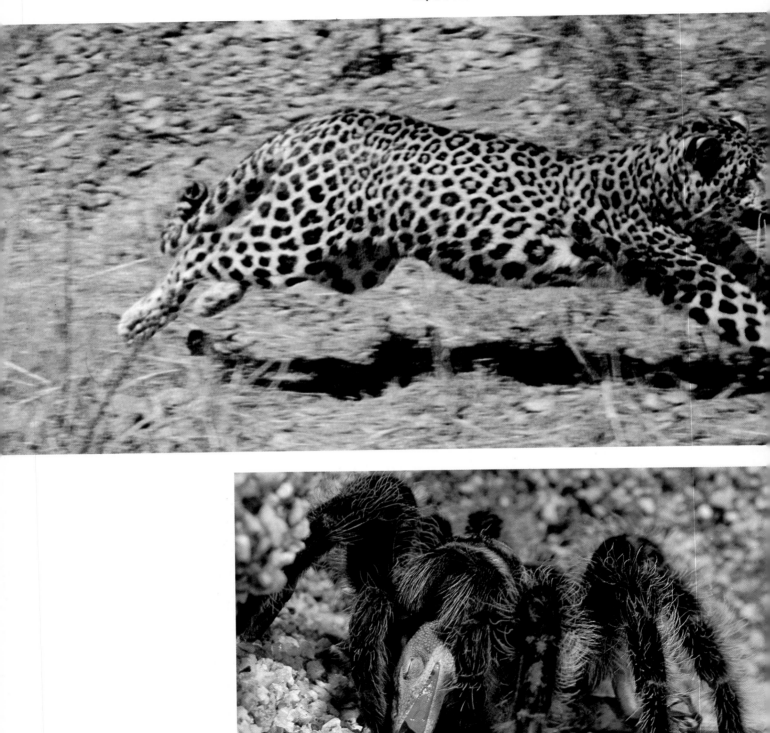

Somewhere out beyond the jabber of the monkey troop, far from the warmth of the prairie dog burrow, the non-cooperators go it alone. Sharing life's risks holds no allure for them; each has worked out another way. Except for the necessity of mating, the grizzly and the polar bear, the cheetah and the jaguar, and many others seem to live in either a shy or a hostile avoidance of other animals.

Yet their capacity for solitude seems to have limits. Males of both the big bear and cat families are sometimes seen mingling with their own kind for a short time. Then they may disperse back to their well marked territories, each to respect the landmarks—the low mounds of dirt and vegetation, the rubbing posts—of others, and to deal harshly with trespassers on its own turf.

Many female loners have the companionship of their young, and their offspring the benefit of motherly protection and training, for a time. But many another—the newly hatched snake, anole, or turtle—begins life truly alone, never seeing the parents that left it the legacy of life. Instinct alone directs them on their way—and unerringly they find it.

No one knows what environmental pressures, what evolutionary imperatives may have combined down through the eons to make some creatures become loners and others turn toward cooperation. But we can see that each way works well and that the variation in their lifestyles makes for more and different biological niches. This in turn ensures fuller use of all the resources of sky, sea, and land and a more secure existence for every creature living wild.

Loners' offspring learn self-sufficiency early in life.

Young bears, lizards, and sea turtles learn quickly that fending for oneself is fraught with hazards. The six-month-old cubs (below) may stay with their nursing mother for two more years, learning from her how to find food and shelter, *to hunt and to orient themselves, and to fear other bears and man. Still, the inexperienced bears will face a precarious life until they are full grown.*

The anole (bottom) and the young sea turtles that will hatch from the eggs just left by the mature Ridley (opposite) know no such schooling. From birth, they must shift for themselves. Birds and snakes *reduce the lizard population considerably before the young acquire survival skills. Gulls, mammals, crabs, or man may snatch the sea turtle eggs or the baby turtles as they inch their way across the sand toward their sea home.*

Sending and Receiving Messages

From whale song to firefly blink, more is being communicated in the wild world than talkative humans know.

The Communicators

One glance is all you need, for the mandrill makes his message clear. To humans and fellow mandrills alike, the bared fangs and riveting stare are a communication not to be taken lightly. Stay away, warns this ferocious baboon, when West African farmers try to shoo him from their croplands. Keep your distance, it insists, or be ready to feel what these sharp teeth can do.

You may never have seen a crimson-nosed mandrill. The arresting photograph on the next page may be the closest you have ever come to one. Yet you can feel from the look on that scowling face the same threat, the same menace, that the photographer sensed directly. Message sent, message received: we call that communication.

We who claim to be the cleverest of the communicators scarcely realize the extent to which animals also play that role. We fail to pick up messages expressed in terms that we do not understand or transmitted by media that we receive poorly or not at all. Yet the natural world is abuzz with messages.

Through sight, sound, smell, and touch animals let their needs and intentions be known. The form of communication varies with the physical abilities of the animal, its particular need, or the intensity of the threat. In one situation, facial and body language might get the point across. Another might call for a change of color, or vocal warnings. Chemical secretions, excretions, even social dances help animals to say what they must say to survive in the complex and demanding wildscape.

The face alone can communicate a whole range of emotions from sexual interest to anger. We humans use that ability well. Our faces are fitted with more separate muscles than any other animal's, and by movements of face and body we can communicate with a repertoire of muscle combinations numbering nearly three-quarters of a million. Each movement has a peculiarly human meaning. When we are happy or feeling friendly, we smile—though our smiles can express a variety of emotions.

In the animal world, too, different smiles have different meanings. A chimp named Ham was smiling after riding a rocket to the void and back in the early years of space flight. What humans saw as an ear-to-ear grin of pleasure was actually an expression of fear. Like the mandrill when he felt threatened, Ham bared the teeth that are the main defensive weapon of the primate. Yet a moderate smile may signal his submission to a neighbor a rung or two higher on the group's social ladder.

The fox or wolf may put its ears into the message; laid back flat, ears reveal a lot of fear along with the submission. Not so with the horse; in equine social circles a pair of ears laid flat back against the head is a sure semaphore of aggression. Yet variations in the way different animals express themselves seldom cause trouble because, except in unavoidable confrontations, animals ignore messages from other species and heed only their own kind.

Other facial parts send messages too. An elephant speaks volumes with its trunk. A caressing trunk shows friendliness; a curled-up trunk means anger or frustration. A subtle gesture, such as an elephant cow using it to nudge her young along, often gets the message across.

A female howler monkey in the forests of tropical America signals her willingness to breed by the way she sticks out and pulls in her tongue. The "courted" male accepts her invitation by mimicking her motions. A male frog puffs up his throat sacs and "jug-o-rums" to entice a mate. But some messages are too urgent to be veiled in subtlety. When danger approaches a herd of elk, the barking or sharp explosive snort of the bull resounds across a mountain meadow.

Some animals use gestures and bodily postures to send signals. Birds show readiness to mate with intricate dances or flashes of plumage. Gorillas signal submission by folding their arms. Chimpanzees indicate peaceful intentions by touching hands, embracing, or breaking twigs. As a friendly greeting, they nuzzle and kiss. When minor conflict occurs, the dominant male gorilla steps in to reassure subordinates with a pat on the back; to settle quarrels between females he becomes rigid and glares at the perpetrator. Extremely dangerous situations, involving the security of self, mate, or young, call for stern measures—violent physical contact to frighten away or subdue the attacker.

A mandrill telegraphs feelings with face and body language.

Human observers find the male mandrill's piercing eyes and enormous teeth disquieting, even when he only lifts his lips slightly in a friendly greeting or to show well-being. But when threatened, his hard stare, growl, and yawning dental display frighten any intruder—man or beast. If a threat persists, the mandrill's red face becomes more brilliant, his chest turns blue, and red dots appear on his wrists and ankles. In a rage, he spreads his arms, slaps the ground with one hand, and throws sticks, stones, and clods of earth. When his tactics fail to scare away the aggressor, he will finally attack.

Previous page: Red Foxes—McNeil River Game Sanctuary, Alaska.

122

Through touch, mammals show that they belong to one another.

Kissing, touching, and grooming give animals a close-knit feeling. Prairie dog pups (top) embrace to coax an adult to play or to groom them. After kissing, prairie dogs may lie down to run paws and teeth through each other's fur.

Cuddling and grooming give rhesus monkeys (opposite) the same secure feeling of belonging. While picking out dirt, dry skin, and ticks, the monkeys soothe away tensions and reinforce family ties.

A female elephant uses her trunk as an affectionate mother's teaching tool: a sharp spank teaches the young to avoid danger, a caress rewards, or (above) a gentle nudge tells baby it's time to leave the waterhole.

A few creatures light up the dark with messages. Light-producing organs in the firefly's abdomen hold chemicals that combine to produce a glow in our fields and gardens. Some species blink their cold chemical light only to seek out a mate. To signal a female of his own kind the male turns his light on and off in a precisely timed and coded sequence. The female acknowledges his advances with matching coded blinks and soon he is at her side—or in her gut, if she turns out to be one of the predatory fireflies that decoy unsuspecting males.

In the sea, too, messages are sent with luminescent signals. At depths of half a mile about two-thirds of the fish species are capable of throwing off light with cells called photophores. Some use the light to find and attract a mate. Viperfish and anglerfish use it to deceive. Dangling luminescent lures, they snap up the smaller fish that dart in to investigate.

Colors also have specific meanings beneath the waves. The octopus uses them to startle and scare away would-be attackers. The normally dull speckled, green-brown reef octopus can quick-change into an arresting bright red with yellowish-white spots.

The secret of the octopus's transformation lies in its chromatophores—large, round cells in the skin that contain granules of pigment. When the animal's nervous system expands these cells, the pigments show. An octopus can open all or a fraction of its pigment cells, enabling it to produce a shade to suit the action—attack, alarm, defense, camouflage, or rest. When the muscle fibers contract, the cells close and the octopus again wears its drab coat.

A discus fish uses its silvery lemon color to signal to cleaner fishes that it needs help in ridding itself of fungus infections. When the "cleaning crew" approaches, the discus fish turns almost black, making it easier for the tiny fish to find the white fungus spots.

More subtle color and texture changes camouflage the flounder lying on the ocean floor. Its chiaroscuro blotches or geometric patterns blend harmoniously with mud, gravel, sand, seaweed, or broken coral or shells. This cryptic disguise works well: it communicates to the flounder's prey "there is nobody here" seconds before the flounder pounces.

With its mouth wide open as if to allow a bellow to erupt, a hippopotamus sends the nearby world a silent, but strong, visual message: stay out of my pool! On shore the hippo has already built a fence—not of mud or brush or stones but of scent. There in the underbrush it flailed at its own droppings with its tail, scattering them at nose level in the shrubbery. In this way it staked out a territory, and with tusk and bluster it defends its claim by confronting intruders with an ominous snort. Sight, scent, sound—the hippo's territorial message is broadcast on all three channels.

The bellows of a male alligator echo through swamp and bayou to invite a mate. Soon a female, looking like a scaly torpedo as she draws a V across the dark water, responds to his call. They mate and the female soon builds a compost nest on the shore to cover her eggs, and guards it from all intruders. Several weeks later, communication of a different sort falls upon her attentive ear: soft chirpings from inside the eggs announce the readiness of the young to hatch. Quickly she digs the eggs out of the nest and assists in their safe hatching.

Such innocent communication can be risky, especially when the peeps come from newly-hatched birds begging for food. Their chirps may attract a hungry predator to an unguarded nest. But nature has equipped the tiny chicks with a high-pitched peep that is surprisingly difficult to trace to its source.

Thin piercing notes serve adult birds as well. The high-pitched alarm call of a prairie chicken can warn its companions of a passing hawk without giving the hawk a signal to home in on.

Even so, an alarm call involves more risk to the caller than would simply hiding in silence. Why then does a bird expose itself needlessly to danger? And why does the prairie dog jerk skyward in its stylized "jump-yip," or warn all its neighbors with its shrill whistle, instead of just zipping into the cool, dark security of the burrow? Perhaps mutuality is the answer. Except when momentarily distracted, animals are always on the lookout for predators. When danger approaches, the first one to notice sounds the alarm. The payoff comes when the prairie dog saved by a warning call today sounds the alarm to save others tomorrow.

A song, a hiss, or a yip warns trespassers to stay away.

Wide-open jaws and beaks usually protest territorial intrusions at top volume. But the big hippo sometimes threatens silently (opposite), his voiceless gape speaking louder than his snort. He reserves the visual threat for strangers that come too close—a last-ditch warning before he attacks.

The motmot of Tobago (top) sings his protest to avoid having to fight over turf, while a crocodile (above) threatens trespassers with sharp teeth and an ominous hiss.

When a prairie dog (bottom) postures stiffly and calls out, his clan echoes his claim: "This land is ours."

129

Dolphins appear to converse with squeals, whistles, and grunts.

Since the white-sided dolphins (opposite) take turns making vocal sounds, human observers assume they are conversing. Their "language," which varies in both pitch and rate of repetition, appears to be learned in the 18-month suckling period by imitating sounds the mother makes.

Because dolphins are fast swimmers and spend 98 percent of their time underwater, studying their language in the wild—with precision—is almost impossible. So marine biologists attempting to decode their "conversation" work chiefly with captive animals. Scientists have identified some 18 variations of the whistling sounds which seem to announce the location or identity of individual dolphins or even to express emotions; however, they are still attempting to match the dolphin's sounds with specific behavior.

Many animals use specific calls to communicate different messages within the family. The ground squirrel chirps one danger call for a flying predator, another for a dog or coyote. When a snake comes on the scene, the squirrel will trill a third variation on its basic theme, and then do an astonishing thing: it sidles up to the snake and holds high its tail, signaling to its fellows, "Look, everybody! Here's the enemy, right here!"

Many birds too have more than one call to indicate degree of the alarm and to identify the intruder. The hazel grouse uses four warning notes to alert its young to different kinds of danger. Although alarm calls may vary according to species and the circumstances, often different species sharing the same habitat have developed strikingly similar alarm calls. One grouse may spot a predator, call out, and the whole bird neighborhood will flit for safety.

Some of the buzzing insects—the cricket, katydid, cicada—have specific mate-location calls. On a summer's day you can hear the cicadas' high-pitched pulsating buzz a quarter of a mile away. No insect rivals them in sheer loudness. Yet the racket of its courtship call—a rasp of wing on wing—is music to another cicada's "ears."

Less is known about the specific calls of the popular sea mammal, the dolphin. Dolphins appear to be communicating when they twitter, yap, bark, squeal, or whistle and bubble air through their blowholes. Sick or injured dolphins may whistle with rising and falling inflection, but such "distress calls" are heard also when no discernible distress exists. Scientists are still trying to decipher the cryptic dolphin "language."

The "music" of whales also mystifies biologists. Evidence so far confirms that the male sings during the breeding season and, like the songbirds, may be identifying his species, sex, and location, or be announcing his willingness to mate and to contest other males.

Whale songs reverberate through perhaps hundreds of miles of ocean, none excelling in virtuosity the song of the humpback whale. Though it only whistles, grunts, and moans, we call its repertoire song, so impressed are we with the musical texture and structure. Like

Scent helps wild ones to claim mates, pups, or territory.

The feathery feelers of the male luna moth serve him efficiently as a nose (below). At mating time, they seek out a special aroma—the female scent. His sensitive antennae can pick up the inviting "perfume" as far as a mile away.

A female sea lion licks her newborn clean, then immediately sniffs its breath (opposite, right). From that moment, mother and pup recognize each other by scent. Later, when the cow returns from feeding in the sea, she will call to her baby and rush to the answering voice. But, before feeding it, she puts the pup to the ultimate test—a good sniff to be sure it is really her own.

The male red wolf (opposite) uses scent to mark off his territory. Males regularly anoint weeds and bushes, especially those around the family den, with urine markers to let other wolves know they have staked their claim.

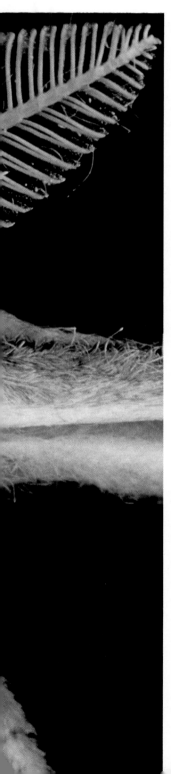

residents in different regions of the United States, humpback whales plying different waters have regional twangs and drawls. Happily, like regional Americans, all members of the species seem to communicate despite differences in their accents.

Performing on his submarine stage, the humpback contributes the longest, slowest, loudest counterpoint recorded in all of nature's symphony. Its versions vary, Atlantic from Pacific; its themes change progressively from year to year. Within about four years, the old song has been replaced, bit by bit, by a completely new melody. Sounding, breaching, sounding again, the great whale sings the current form and theme, then rests and replenishes its 30-ton hulk with a feast of tiny krill until it's ready to belt out an encore.

In the wild another sense—the sense of smell—plays a more important role than humans realize, for we are more attuned to sight and sound. Seal mothers smell their newborns' breath and then depend upon memory to identify their young for life. Land mammals recognize territories fenced with scent posts—glandular secretions of such animals as deer and beaver, or urine of dog and wolf and mouse. Even in captivity this kind of signaling persists. Zoo deer sometimes rub their head glands against a familiar keeper to mark him as their own. And a mouse whose cage is cleaned too often may actually die of dehydration as it strains to mark permanently with urine its unscented new territory.

Within its cloven hoof, the whitetail deer carries a special gland charged with pheromones that help deer track one another and may communicate danger. As a frightened buck bounds away to safety, the gland scents his hoofprints with warning signals that will alert other deer. And he reinforces the scent message with a visual one: his upraised tail.

Insects too depend heavily on chemical signals. Despite the deafening volume of a few insects on a summer's night, most species make no use at all of sound. The male luna moth can locate a female by following her scent on the breeze, even if she is a mile away. Plumelike antennae crown the head of the male, each antenna almost literally a sieve for straining

from the air the faintest trace of a female phero-mone. His antennae can detect a single molecule of the female's come-hither chemical! All that the female has to do is to release some of her "perfume" and wait. But timing is impor-tant, for the male cannot fly against a hefty headwind. If the release coincides with a breeze, he can use the pheromone as his homing bea-con and climb up the wind to reach her.

A fire ant that has found food also leaves a silent trail of chemicals when it reports back to the nest. Others follow the trail to find the food. Had the scout encountered dan-ger instead, it would have released alarm pher-omones to alert the entire colony to it.

The ability to send and receive messages plays such a vital role in the close encounters of the social animals that they use every medium available to them. To make its meaning clear, the anole uses body language that amounts to a visual shout. The bright-hued flap of hide bal-looning beneath the small lizard's throat does double duty: to other males it serves notice to stay off his turf during mating season. But to the female of its kind, the signal serves as a lure; to her it is love talk, an invitation to mate. Thus messages flashed by the anole's stretchy dewlap contribute to the mating pro-cess and thereby to survival of the species.

The versatility of animals in using multi-media language is impressive enough, but when they want to attract a mate their com-munication skills take on real flair and novelty. Before an animal reaches maturity, its mes-sages are usually straightforward and clear: to threaten a bite, show some teeth; to show sub-mission, retreat. But courtship communication takes on forms that are not always as easy for us to decode. Ms. Anole, whatever do you see in that puffed-up flap of skin?

The naturalist calls such physical features a mating badge. Many creatures have at least one to show off when it's time to mate. The buck deer bears a handsome rack on his head. Many birds flaunt nuptial plumage; some strike acrobatic poses to display their feathers to best advantage—a bird of paradise hangs upside down to show his colors. Perhaps the most dramatic of the birds is the peacock, whose timing would match the best of theatrical

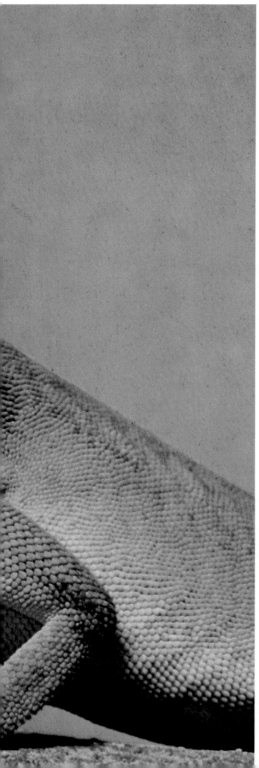

A puffed-up body speaks volumes to a potential predator or mate. The anole's showoff dewlap (left) helps him in two ways—it repels attackers and attracts mates. With the same ploy, a frigate bird (right) attracts nearby females and frames his distended throat sac with impressive wings that stretch out up to eight feet.

Frogs puff up their vocal sacs (above) when they croak out a love song. As their mating call ends, they exhale and the "cheeks" collapse. Wild turkeys from Texas hill country (opposite, top) also combine voice with a fine display. They flash tail-feather fans and fluff up body feathers as they gobble and strut to win a hen.

troupers. When approaching a peahen he struts on a diagonal course, with the drab underside of his fanned-out tail facing her. Then, just before he reaches his intended, he abruptly pivots around to bewitch her with the dazzling color of the "eyes" on the upper side of his quivering fan. To leave nothing in doubt, he climaxes this show with a piercing cry. Other birds, lacking attention-getting feathers, advertise sexual readiness with colorful wattles, combs, or feet. The male blue-footed booby will parade in front of the female, feet held high to attract her admiration.

Mating badges are found in all life forms. A male jumping spider has a white stripe bounded by iridescence on his abdomen which he shows off to a potential mate by zigzagging up to her and waving his legs up and down. The male cuckoo wrasse changes color for one month each year in the spring. From a muted yellow and blue, the fish changes dramatically to gaudy gold-blue patterns with white splotches on its neck and face. In some fish species, both sexes change color. The female jewel fish shows red mating colors as bright as

*"Come dance with me" means
"be my mate and share my life."*

*Normally stately, slow-moving
cranes and storks cavort in
exuberance when they dance.
The courtship ritual of the
Japanese red-crested cranes
begins when the male bird
bows to his partner (top left),
hops toward and around her
with wings outspread (top
right). The birds perform the
same sequence of steps, either
together or alternately. After
they mate, one or both birds
may resume dancing.*

*The dance, which usually
takes place in late winter,
lasts two or three minutes
and is repeated two or three
times in one day.*

*In contrast to the crane's
gentle poses and postures, a
pair of yellow-billed storks
(opposite) use the same movements that males use in fighting—with a difference. When
male and female flap their
wings and hit bills together,
they are getting ready to mate.*

the male's, sometimes causing inexperienced
younger fish to chase the wrong sex.

Mating communication reaches its aesthetic
height, many bird-watchers agree, in the dance
ritual of cranes. The dance movements vary
from species to species, but all are graceful,
enchanting pas de deux. With outspread wings,
the tall birds circle and leap stiffly several
feet in the air, occasionally bowing low to
the partner. From time to time the birds may
fling twigs into the air with their beaks. Some
birds call out during the ritual; others perform
silently. The dance ends with a wild flapping
of wings and a backward snapping of the head.
The birds then resume their normal erect
posture with head aloft.

Biologists do not know the roots of these
urges and rhythmic gestures. But they do know
that without a dancing "partner" a female
crane will hatch no young. The dancing ritual
synchronizes the male and female sexually,
making mating and fertilization of the eggs
possible.

At each stage of its life an animal feels the
need for some bit of territory it can call its
own. Even in flocks, birds usually allow each
other a certain amount of space, as you can
see by the gaps between birds roosting on
a telephone line. In order to mate and to bear
young, birds and other creatures must give up
some of their individual space. Many pairs, for
instance, must share both shelter and food for a
time at least. A courtship ritual eases this ab-
solutely necessary transition from solitary to
communal living. Yet movements in the ritual
often resemble attack gestures, so the male
sings and uses precise postures and movements to reassure the female that his intentions
are friendly. By repeating rituals until the female's instinctive urge to defend herself sub-
sides, the male succeeds in winning her confi-
dence, closes in, and they mate.

One mating ritual that is touching to humans
is the gift-bearing of the blue-footed booby
who brings a bit of greenery in his beak. Some
other birds swap nesting material or offer
gifts of food. Communication through such
rituals not only encourages peaceful sharing,
but builds the bonds that are necessary if the
species is to reproduce.

Like humans, animals get diseases, have accidents, and suffer injuries. But often they do not show their misery, for animals have a high threshold of pain and thus do not display discomfort until their condition is severe. A hunter dressing out a healthy-looking moose may be surprised to find its lungs infested with parasites. A live deer or catfish may seem unaware of a gash on its flank.

The reason is quite basic to survival, for predators are ever watchful for the slightest trace of weakness in their prey. Signs of sickness invite attack—slower-than-normal movements, limping, bleeding, reluctance to counterattack—all say to a predator: you are stronger, the odds are in your favor. And even these signs can be misread: the shark sensing the vibrations of a human swimmer hears not the slippery grace of a healthy fish but a thrashing more like an injured one.

Communication style and sign may vary, but most messages get across. Yet even within a species a dispute may lead to a fight. A male giraffe asserting dominance lowers its head and neck horizontally. If his opponent does not submit by raising its nose in the air, the aggressor will start a head-slamming contest—blows against each other's necks, followed, if necessary, by stronger blows to the neck and body.

Messages sent by one species to another are more frequently not received correctly. Animals, like humans, often get entangled in a "communication breakdown." When that happens, an attempt to inflict pain becomes the ultimate communication medium.

Under threat of attack, the porcupine first erects its quills. Then it turns its back to the enemy, tucks its head between its front legs, stamps its feet, and rattles its quills—fair warning to the fisher, bobcat, lynx, or coyote that tries to corner what it regards as a tasty meal. If the attacker disregards the porcupine's forceful communication, it may live to regret it. Lunging at the predator, the porcupine's last word is to jab the enemy with its quill-laden tail. Pain and regret follow for the predator that was not on the right wavelength with the porcupine. A rattlesnake's response to threat similarly escalates. If an intruder persists, a soft warning rattle is followed by hissing and vigorous rattling before the snake poises its head and strikes.

Of all the ways to get a message across, the surest—and most risky—is fighting. Biting, butting, battering drives a message home. But animals learn early that violence wastes energy, that it can weaken, wound, or kill. To avoid fatal risks, most animals fight as a last resort after having explored all the alternatives in their repertoire of messages. By so doing, wild creatures demonstrate the real purpose of communication: it makes togetherness work.

When warnings fail, fighting becomes the last word.

Fighting seldom erupts spontaneously, but often follows when threats are persistently ignored. The snarling hyena baring its teeth at a wild dog (opposite, top) may have been trying to snatch a pup or some meat from the dog pack's kill.

Rival rams habitually duel over breeding ewes (opposite, bottom). A hoof jab and a grunt invite a swift charge. The bighorn less hurt by the clashes claims the ewes while the dazed, more bloodied loser retreats.

Attempts by a male giraffe to establish physical and sexual dominance in a herd may escalate from head-slamming (below) to more serious fighting. Giraffes then stand close to- gether with stiff legs splayed and swing their horns against the opponent's neck, shoulders, sides, loins, and flanks. Head blows may knock a giraffe unconscious. The weaker male eventually withdraws and the winner pursues him briefly to seal his victory.

Passing on the Secrets of Survival

To bear the young and
to ensure their survival is
life's ultimate purpose
for every species.

The Courtship Strategists

As breadwinner and defender, an animal grows to maturity. As communicator and cooperator, it finds a place among its kin and kind. As builder and traveler, it strikes its own truce with an unyielding environment. Whatever its species, the reward each creature receives for playing these roles well is its own survival. But a time comes when the animal must play out a role that is of little direct benefit to itself. It must give of its substance, its energy, and perhaps its very life, in the replication of its kind. It must find a mate and bring forth young.

For many creatures, this is a time of great peril. The safety of hiding or camouflage may have to be forfeited in the search for a mate. The energy spent in courtship ritual may suddenly be needed for defense. Food must be shared, territory defended, rivals repulsed, bodily changes endured, mobility sacrificed. Because the survival of the species depends upon animals undertaking the burdens of courtship and parenthood, nature instills in virtually every creature a drive as inexorable as hunger and thirst. Each in its season is driven to mate.

There are exceptions, a few creatures who need not mate to reproduce themselves. Some parameciums, slipper-shaped specks of life barely visible to the human eye, can simply multiply by dividing; where there was one, now there are two. More than two dozen kinds of reptiles and at least one kind of fish have no male of the species, or almost none; each female begets daughters exactly like herself.

Why then does nature bother with sex? Why do with two what can be done with one? The answer lies in the need for variety. Earth is a changing planet, a lab whose experiments do not always allow for creatures that are slow to change. The prize of survival more often goes to the adaptable ones, and mating—as a stirring of the gene pool—plays a vital part in achieving adaptability.

The drama takes place in the secret world of the living cell. There lie the chromosomes, coded like computer tapes with the chemical commands of the genes. Each gene dictates a characteristic; in combination with genes from a mate, they bestow dark fur to this wolf and light fur to that, a bit more speed to one salmon, a bit more endurance to another. But a parent without a mate gives its offspring no genes but its own, and so the generations peel away like endless carbon copies. Except for an occasional mutation, they vary only gradually as the genes themselves change. Each union of a male and a female shuffles their differing genetic decks and deals out a whole new hand for their progeny. And even the paramecium plays the game occasionally; when times get tough, pairs sidle up to each other for a swap of genetic material in order to survive.

Of all the roles played by nature's creatures, none is as complex as the mating game. It must be synchronized precisely, for if one partner reaches readiness before the other, the efforts of both will likely be wasted. It needs to be tuned to the environment, so the young will emerge at the best time and place for their survival. And it must join the right individuals, of the right sex and species, at the right spot and moment—with a minimum of risk and effort.

For no two species is the scenario exactly the same. The differences begin with how each recognizes or senses the proper time. An internal calendar seems to call the maturing salmon back to its natal stream to spawn. For the female mouse, the reproductive trigger is the scent of the urine of the male. Some insects mate only when certain proteins appear in their food.

For most vertebrates, and many other creatures as well, day length seems to be the trigger. Light enters the eyes or skull of a bird, registers on the forepart of the brain, and is relayed by chemical signals to the pituitary gland. The pituitary prods the gonads, they in turn flood the system with hormones, the hormones trigger reproductive behavior—and all outdoors is regaled once again with the color and clamor of the courtship of the birds.

For some insects a scent trigger is all that's needed. A male butterfly sensing a female pheromone on the breeze simply homes in on it and mates with its source. No need for a courtship ritual to verify her species, sex, or readiness to mate; the pheromone says it all. The male fertilizes her and departs; soon the female lays her eggs, and she too goes her way.

In a race with time, marathon swimmers rush to spawn.

After wandering for two to three years in the Pacific, the urge to mate and reproduce drives sockeye salmon back into fresh water.

Hormonal changes tell salmon when to start their long journey and a built-in compass, that may use the sun, moon, and stars or the earth's magnetism, shows them the way. With such navigational aids, they chart a sure course toward their home river, then use their keen sense of smell to remember the odors of the feeder stream where they developed and even the rivulet where they hatched.

All salmon undergo some color changes, usually at the time they enter fresh water. The sockeyes change spectacularly to a crimson with an olive-green head and the male acquires his mating badge— a humped back and an elongated hooked nose, shaped to ward off other suitors.

Previous page: Cheetahs—Serengeti National Park, Tanzania.

But courtship is rarely that simple. For many other species, the place as well as the time must be right. Some slugs mate at the end of a thread of mucus hanging from a twig. Honeybees seem to mate in midair; the queen and her drones fly from the hive, and when she returns, she is fertile for life. In some species of anglerfish, the male's idea of mating territory may be the most bizarre of all: he simply attaches himself to the first female he meets. His tiny body fuses to that of his much larger mate, and there he dwells as a parasite for the rest of his life, slowly degenerating to little more than a shapeless sperm bank. In the stygian gloom of the ocean deep, it is chancy enough to find a female even once; by this grotesque ploy the species ensures that he won't have to do it again.

To announce its selection of a place to breed, the blackbird calls from a cattail, the ruffed grouse "drums" on the air, the woodpecker beats a tattoo on an oak. At the same time, each lures a mate, warns a rival—and risks a predator's attack. The northern oriole sings his territorial song in suburbia where insects rustle in the shrubs and gardens, because prospective mates will judge him largely on how well his spread will feed a brood.

Among creatures adapted to abundance, territory is often a permanent and practical matter. But where males (or females) are more

144

Song and game birds announce claims to a breeding place.

Deliberately making themselves conspicuous, birds of swamp and forest let their neighbors know decisively that trespassers are not welcome in their chosen breeding spot.

The yellow-headed blackbird (opposite) whistles and squeals a harsh, rasping challenge to let males of all species know that he'll battle to protect his ground. In reedy lakes and marshes, this aquatic specialist chooses a compact, easy-to-defend area close to his main food supply of water-borne insects.

The ruffed grouse (left) claims a patch of forest not with a call, but with a beating of wings. Perching usually on a hollow log, he raises his ruff—neck feathers—perhaps to look larger, and starts drumming. As he fans his wings up, forward, and down about 40 times a second, the cupped air resounds in a rhythmic thump through the dense woods.

highly competitive and more promiscuous in their mating, a territory may be merely symbolic. Animals as diverse as the prairie chicken and the Uganda kob win mates by fighting for strutting spots in a courting arena called a lek. The fighting demonstrates the strength they will pass on to their offspring and the females pick the winners.

But when mating ends, all claims to the lek expire; the males abandon the sites they fought so well to win, and the females bustle off to rear the young alone. For the prairie chicken and the kob that tiny patch of territory has become useless. All that mattered was the ability to win it.

Some of the fiercest courtship fights focus not on territory at all but simply on the right to mate. Male elephant seals climax their spring migration to California's offshore islands wit furious fighting as soon as they hit the beache Heaving their ungainly bulks ashore, the bull launch a raucous melee for dominance. Lou snorts originating in the bulbous nose forc sound waves into the mouth which serves a a resonating chamber. Canine teeth strike at rival's nose and gouge from his body bi chunks of the blubber that constitutes one thir of the animal's weight.

By the time the females arrive, the season social hierarchy has been established. Th alpha bull may mate with more than half of th cows and a few other victorious males will g most of the other half. Each cow is bred soon as she gives birth to the pup she co ceived a year ago. Even as one of her two ute uses begins to recover from this pregnanc

the second receives the egg and sperm for the next one. But until the last cow is bred, high-ranking bulls must hold off lower-caste bachelors trying to improve their status.

The resounding clash of elk antlers climaxes a similar attempt by a bachelor elk to steal a cow from a harem master. The older bull does not try to guard a specific grazing ground, but frantically patrols the harem he has rounded up from a summer herd of cows and calves. Besides dealing with interlopers, he has to fight to keep out the yearling males he separated from their mothers—and to keep in the harem the cows who want to escape.

When another male intrudes, the harem master tries to rout the rival by threatening him with sharp antlers and by engaging him in head-to-head pushing matches. But when the master realizes he faces a serious contender, the showdown takes on a duel-like formality. With neck outstretched and rack laid back, the master goes to meet the invader. The challenger bugles and slashes the air with his antlers. Advancing on a collision course, the bulls halt about 15 yards apart to circle each other before making the first of several head-on charges. Antlers intertwined, hooves dug in, each attempts to twist the other's head enough to throw him off balance. Then one quick broadside attack with the pointed rack will knock the other down. The duel is over and the loser retires—to raid another harem or to face another year of celibacy.

When the autumn rut ends, the rack has no further use and is soon shed. Unless, as sometimes happens, it has already become a deadly liability. It is a sad end indeed when two lordly elk who locked antlers in a courtship joust find they cannot free themselves. Starvation is their fate, their only memorial a pair of skulls with the antlers still enmeshed.

The courtship fight itself can be fatal for golden pheasants. The onset of warm weather in the Chinese highlands and the appearance of a single hen will incite two cocks to prove who is more fit to mate. With incredible speed, the swains hurl themselves into a flapping, pecking, spurring brawl that can leave one bird dead and the other so battered that he is barely able to mate with the waiting hen.

147

Singer, strutter, and interior decorator woo their mates.

When courting a mate, animals draw on their unique resources.

The ornate peacock (opposite) relies upon his iridescent feathers to bewitch a female. All he need do to win her is fan out his feathers to show her their lustrous "eyes."

The less-endowed tree frog (left) makes the most of the attention-getting throat pouch that amplifies his mating call. Only females of his species can tune in and respond.

To make up for plain looks, the great grey bower bird (below) decorates his inviting bachelor pad with ornaments and blossoms. When an attracted female hops inside, he flutters about wildly, seducing her with his treasures.

Fortunately for more docile species, fighting is not the only means of winning a mate. The resplendent peacock does it purely by strutting. Though he bears himself like the veriest egotist before his drab little peahen, he is really a supplicant. Having reached maturity, he now feels the full fervor of the mating instinct. But he will not mate unless she accepts him, so he deploys his gorgeous plumes to that end. It is fitting that he puts on such a splendid show, for she will get little else from him. The mating game is often thus: the more impressive the male, the weaker the bond between the partners.

Few court more ingeniously than the bower birds of Australia and New Guinea. To mate with as many females as possible, each male not only builds a special trysting place, but decorates it with trinkets. One of the 18 kinds of bower birds, the crestless gardener, looks exactly like his dull brown mate—so he compensates by building the gaudiest bower, a charming little stage with an overhanging roof and a decor of berries, pebbles, and fresh flowers. Other bower birds build maypoles, using a sapling as the centerpole of a conical hut or transforming its trunk into a decorated column ringed with moss. Those that collect trinkets may steal them from rivals—and wreck the rival's bower for good measure. The array may include pebbles, shards of glass, coins, a shiny camp spoon, a metal tab from a soft drink can. Some of the bower birds are artists; they mix green or blue plant juices with saliva and paint their bowers with their beaks. The satin bower bird qualifies as a tool user as well, for it daubs charcoal and saliva on its walls with a wad of bark.

By signal, lure, or circumstance, the sexes meet. But even then, many courtship strategists still have several questions to be answered. The first is: are you the right species? There are many kinds of fruit flies, but in each species the male waves his wings in a pattern unique to his kind. And the females respond only to the pattern of their species; any other sends them flying.

The next question is: are you the right sex? Not even pigeons can always tell pigeons apart. The male struts and displays to any pigeon

149

Couples achieve readiness to mate through precise rituals.

Animals' playful courtship rituals reveal sexual interest and lead to acceptance of mates.

A male green orb spider (top) hovers over his mate to stroke her with vibrating front legs; if the female is receptive, she tucks her legs in.

To show her acceptance of a male's advances, a female western grebe joins him in a water ballet (opposite, top).

The love rituals of lions (above) and masked boobies (opposite, bottom) look more like squabbles. Lions play a swatting game and boobies clack beaks to test mutual attraction. If the chemistry is right, the match is on.

handy; if he gets an aggressive male response, he moves on to another.

Much of courtship's energy—and ingenuity —focus on the next question: are you ready to mate? The male scorpionfly sidesteps this one by presenting a female with a freshly caught fly, then copulating while she is preoccupied with eating it. One species of dance fly embellishes the fly with a little silk. A third presents a silken package, but the gift inside is much smaller. And a fourth gives a package with nothing in it at all.

A male spider that comes across a female's web or dragline can probably answer the questions of species and sex by the smell or taste of her silk. Now comes the question of willingness: are you willing to mate with me? For the nearly blind spider, it is a fateful question as he approaches the huge female, also nearly blind and pre-programmed to seize and kill anything of suitable size, including him. So he taps the web gently in a precise code. If she is ready to mate and is willing to receive him, she plucks the web with her own come-hither rhythm; he enters warily, and they mate. Here the male

enjoys an important bit of anatomical life insurance. Male spiders are among the very few creatures that make sperm in one section of the body and deliver it from another. Leglike mouthparts called pedipalps pass the sperm to a pocket in the female's body without a close coupling that could suddenly turn lethal. For their size and function the pedipalps are incredibly complex; most will fit one species of female and no other.

In most courtships, the creatures' best senses are used. The hunting spiders have good vision, so males of these species have bright color patterns that are displayed to the female in droll little dances. Some can even flash their eyes by shifting the pigment around.

Among the higher animals the questions of readiness and willingness may not be so promptly answered. Indeed, one of the functions of an elaborate courtship ritual is to take all the time necessary to ensure that all the answers are affirmative. The ludicrous aquacade of the western grebe is no laughing matter to the courting birds, for it helps them to reach readiness together. It also increases the chances that they will stay together for as long as the young grebes need the care of two parents. For many animals that is the clincher question: will you help raise the young?

For days or even weeks, the pair act out rituals, some of which foreshadow the parent role. Female birds sometimes beg for food like baby birds; often the male responds by regurgitating an offering. Or the male bird may present a prospective mate with a symbolic shred of nesting material before a real nest is needed. In effect, he is making a promise. "Mate with me," the suitor seems to say, "for I know you will need a nest and I will help you to build it." The bond between the pair is strengthened by this promise, and seldom does it go unkept. Indeed, it has become a part of the courtship ritual precisely because the building of a nest is important to the suitor; this alone will guarantee his help. Like many elements of courtship, the gesture prefigures the parent role ahead. And a symbolic twig may be all it takes.

It takes more than that, however, to win the fussy heart of the female house wren. The male

*Even when nest building is used
to draw mates together, there
are no fixed rules about who
does the work. The crowned
cranes of Africa (opposite, left)
and the more widespread cattle
egrets (opposite, right), like*

*many birds that mate for life,
share the task.*

*The crane chicks will hatch
in a soft, grassy depression,
while the baby egrets will
nestle in a bed of local plant
material assembled in a tree,
bush, or among reeds.*

*But among Siamese fighting
fish (below) the male alone
builds the nest. To construct it,
he blows bubbles through
his mouth which secretes a*

*substance that holds the bubbles
together. When a mature
female is attracted to the nest,
she lays eggs below it; the male
fertilizes the eggs, then catches
them in his mouth and blows
them up into the nest. In the
same way he later retrieves
escaping hatchlings.*

who woos her must arrive at the breeding
grounds well ahead of her in the spring. There
the frenetic little bird stakes out a territory and
routs intruders with a lion-hearted aggressive-
ness. His presence fills the backyard with a
tumbling torrent of song that may be repeated
four times a minute. And to the intruder he
sings a different tune, a raspy challenge
fraught with menace.

Some birds may ignore it with impunity. But
when another male house wren intrudes, there
is no forbearance; he is summarily banished,
by force and fury if need be.

Bustling about his territory, the suitor
stuffs every likely cavity with twigs and
debris. Almost any cavity will do—a mailbox,
a beer can, an abandoned hornet's nest. And
almost any nesting material will do for stuffing
it—nails, wire, paper, cloth, string. The industrious male has a virtual housing development ready for the female's inspection. Like a
solicitous real estate salesman, he escorts her
on a tour of his prospective nurseries; like a
finicky prospect, she usually rejects every one
of them. Thus one of nature's stormier romances
begins. First she sets about finding her own
spot. If she does accept one of his choices, her
first move is to tear out his handiwork and build
her own nest from scratch.

The male house wren is not disposed to take
this lightly. Fights between the newlyweds
erupt at every turn as each asserts its opinion
on how to construct a nest and where. Yet even
this is part of the courtship procedure. When it
is completed, the two birds have mated several
times and are committed to the nest and to the
offspring that soon will find life therein. But
not exclusively to each other. After all that hard
work and domestic squabbling, the male house
wren may do it all again, wooing, winning—
and wrangling with—a second mate even while
honoring his commitment to the first.

Commitment: that is the message often con-
veyed in the courtship rituals of species whose
offspring will demand the care of not just one
parent but two. And one of the clearest ways
to signal commitment is by building a nest.
Thus the male flicker pounds and chisels
away at a nest hole in a dead tree trunk so
single-mindedly that the females must begin

the courtship sequence by fighting among themselves over who gets him and his rough-hewn
home. Inside the crude hole the winner then
chisels out a smooth-walled nursery, using the
chips as a soft bed for her eggs.

Softer still is the bubble nest, built, main-
tained, and guarded by another committed
suitor, the male Siamese fighting fish. As soon
as he has blown the sticky bubbles, the female
lays her eggs in the water and he fertilizes them.
Then, before the fertilized eggs can sink, he
carries them up to the floating nest, one by one,
and spits them into the bubbles. There they
will hatch 32 hours later.

Whether of sticks in a tree, grasses in a
marsh, or bubbles on the surface of the water,
a nest bonds together the pair that build it, and
the next generation is the beneficiary.

Water-borne eggs and sperm unite if conditions are right.

External fertilization calls for special techniques to ensure against the loss of eggs. The male European common frog (opposite) clasps a female tightly behind her forelegs. In this way he can deposit his sperm right

on top of her one to two thousand eggs as they are ejected, before they sink to the bottom of the pond.

Brook trout (below) have a more elaborate breeding routine. The female first hollows out a protective nest in the gravel and sand of a rivulet. She then moves to the bottom of the nest to deposit her eggs. Simultaneously, the male arches his body over hers and

immediately discharges his milt (sperm) onto the eggs (bottom) which stick to the pebbles. The female covers the fertilized eggs quickly with loose pebbles and sand before other fish eat them or the current whisks them away.

In the frenzied splashing of a pool full of mating frogs, the males seem to heed one rule only: if it moves and is the right size, clasp it and mate. It's actually a sound rule, for frogs, like trout, fertilize their eggs outside the parents' bodies. No safe, warm conduits guide egg and sperm to their union deep in the security of a womb. Instead they spill out into the water and unite only by luck. Many are lost, so many more must be loosed by the parents than are actually needed.

But in waters aboil with frogs, how does a male tell male from female? He doesn't. But when he grabs another male, a special cry signals the suitor to let go. Only females heavy with eggs fail to sound the release call.

Bullfrog females are choosier. They need only swim among the territorial males and make their choice. The female toad enjoys no such privilege—indeed she may be drowned by over-ardent suitors embracing her two or three at a time.

By comparison, a life launched inside a female animal's body seems pampered indeed. Internal fertilization is an amazing adaptation by land animals of the sea's ancient union of egg and sperm in salt water. Some land-dwelling species have met the challenge of finding and uniting with a partner by endowing one creature with both sexes, a phenomenon called hermaphrodism. The earthworm is at once male and female; it need only find another earthworm of the same species. Another hermaphrodite has developed perhaps the most amazing device at the matchmakers' command: the "love dart" of the European edible land snail. After some warm-up grappling, one snail shoots into its mate a sharp calcareous rod that excites it into accepting the other's sperm. The shot is returned, the swap completed, and the snails separate, each to lay fertilized eggs in the soil.

Love darts, gifts, fights, bowers, songs, dances, postures, odors, colors, nest building— all these and more make up the complex repertoire of courting animals. Each animal faithfully follows its own unique script in the correct order. For even a small omission can break the spell and cost the careless a chance to pass its genes along.

Anatomy enables many animals to mate internally.

Side by side, back to back, or front to back, most land animals join bodies to mate, ensuring a safer union of sperm and eggs than is possible for most aquatic animals.

The earthworm pair (top) come together in a head-to-tail position so that male and female sex organs, which all earthworms have, may deposit eggs and sperm in a special cavity to be held until eggs are laid in the soil.

When Icarus blue butterflies mate (bottom), the male's muscular claspers grip the female for hours as transferring the sperm is a slow process. If disturbed, the pair fly off locked together, one butterfly with wings folded traveling as "passenger."

The male grizzly bear (right) and giant tortoise (opposite) mount to mate. Safe from predators and other hazards, the cub fetus develops within the uterus, while the tortoise eggs are soon dropped into a cache to develop and hatch on their own.

The Parents

It is a perilous journey from birth to maturity. Many trillions of new lives start the journey each day, each hour, each minute. The newborns—whether of one cell or millions—face fearsome odds, and most of them will never reach the goal. And so the fate of any species rests on those that do manage to mature, for they alone can reproduce. Theirs is the role of the parent.

For some species the word "parent" need not be plural; one adult is enough to produce offspring. Streams, large ponds, and lakes teem with a certain kind of water flea—a minuscule crustacean—in whose millions there usually are no males. The females normally reproduce by parthenogenesis: they develop from unfertilized cells. But under adverse conditions, such as overpopulation or changes in water temperature, males are produced. To survive winter, for example, the females bear a few males and then lay eggs that, once fertilized, can withstand the cold. In spring, all these eggs hatch into females that also breed females.

Parthenogenesis, or virgin birth, is a mixed blessing. Because only females are produced, the number of offspring in each generation is roughly double what it would be if half the species were male. Consequently, parthenogenetic species sustain their populations even though hosts of hungry breadwinners eat them. But unisex works only in a steady, dependable environment because, without frequent interactions between the sexes, there are few chances for adapting to change.

Small sea creatures called obelia have developed a more consistent reproductive pattern: one generation reproduces by sex, the next without it. In its plantlike phase, the creature grows a bud from one of its stems, then casts it off as a free-swimming jellyfish called a medusa. But the medusae are male or female, and by eggs and sperm they produce larvae. These moor themselves to the ocean bottom and soon grow into plantlike creatures budding new medusae without benefit of sexual union.

In obelia, as in all animals that reproduce sexually, the sperm is vital. Once the sperm enters the egg cell, the egg begins its plotted growth. For many of earth's inhabitants, that is how new life begins. And for most of them, surprisingly, that is where parenthood ends. The female lays her eggs, the male releases his sperm, and the job is done. The parents never know their offspring.

We who love and care for our children find ourselves in a rather small minority. Ours is the most complex and demanding of the various ways to be a parent. How easy it must be simply to abandon the eggs; no wonder that's the most common method. It's also the most wasteful, since without a guardian parent, many eggs must be laid so that a few may survive. One female codfish lays so many that the Atlantic would be packed solid with codfish in only a few years if all the eggs hatched and the young survived to adulthood.

Some creatures have evolved ways to protect their eggs. The reproductive strategy of sharks and stingrays is far more advanced than that of the cod, for some species fertilize the eggs internally, develop the eggs in a brood chamber, and give live birth. What safer place for eggs than inside a parent's body?

Yet creatures that lack a brood chamber have evolved other ways to safeguard their eggs. A newt wraps her eggs in water-plant leaves and glues the package shut. The female barking frog of southern North America deposits her 50 or so eggs in the moist crevices of the species' rocky habitat and then leaves them. But the male takes over as sentry, guarding the eggs and even protecting them from drying out in rainless times by urinating on them. His vigil may last as long as 35 days before the eggs are safely hatched.

The odds improve dramatically when parents care for both eggs and young. Then birds continue their kind with clutches of fewer than 20 eggs; for the rockhopper penguin, laying only two eggs is plenty. Yet caring for young does not always eliminate the necessity of rearing large broods. The social insects, which deftly protect their young, hatch huge numbers of eggs—the honeybee queen sometimes lays more than her weight in a day—but many are needed to offset the adults' short life span and keep up the nest population. By caring for eggs and young, these insects make sure that little of the colony's reproductive energy is wasted.

Parental chores begin even before birth of the young.

These rockhopper penguins, like many other parent birds, spend hours brooding—sitting on the clutch of eggs to warm the fragile embryos inside. Male and female rockhoppers share this parental duty, the one relieving the other so it can waddle to the water to fish.

In the icy Antarctic clime where these penguins live, eggs must be covered at all times; but in warmer areas, parents have an easier time of it. They may leave the nest to feed as long as the ambient temperature will keep the eggs' temperature from falling too far below the level needed to develop —about 93° F.

Events happen quickly for this female whitetail deer giving birth after 200 to 210 days of gestation. Her contractions force first the fawn's front feet, then its head and shoulders out of the vaginal tract (below). Soon after the fawn is free and the umbilical cord has broken, the doe eats the placental sac—the filmy cover on the fawn—removing it from the sight of predators and gaining nourishment for herself. She also licks the fawn dry (opposite), removing odors and thereby making it harder for predators to find. Very soon, within 30 minutes or so, this doe will be able to lead her fawn—already gorged with its mother's milk—away to a hiding place where it will remain for about two weeks until it can follow her as she browses. Normally a doe drops one fawn in her first pregnancy; afterwards, if she has been well nourished, she has twins.

A honeybee is born pre-programmed; once out of its cell in the comb it needs no instruction as it bustles about its duties instinctively. But for creatures with much to learn, parental care is essential to survival. Newborn water fleas swim blithely—and brainlessly—away while newborn deer begin a year or more of dependence and immaturity. And the more the newborn has to learn, the longer the dependency.

Indeed, even the young deer's mother may have to learn her role. Mammal mothers often appear uncertain, sometimes even fearful the first time they give birth. Instinct seems to have left a gap at this critical juncture, and sometimes the firstborn pay for the omission with their lives.

If all goes well, as it usually does, instinct and observation of her peers guide the new mother in what to do. As the fawn edges toward the udder for its first taste of milk, its mother turns to lick her infant. In this act, the doe does what all mammal mothers do: she forges the first and most enduring link between mother and offspring. The doe's licking also swabs away the enclosing sac and afterbirth, removing the odor that could attract predators.

A fawn can walk within about half an hour after its birth, and its mother leads it to a safe hiding place. But a fawn is incapable at first of the bounding sprint that is a deer's best defense; it relies instead on its wondrously effective camouflage and its lack of telltale odor. A keen-nosed predator may stalk past the motionless little fur ball only a few feet away and never catch a whiff. But as quickly as the fawn's running skill develops, so does its odor. It then begins to travel with the herd and to learn the danger signals by which the others say, "Run!"

Among mammals that must run to survive but lack the protective camouflage that enables the very young to hide, newborns almost seem to hit the ground running. After a few shaky minutes of tottering, a wildebeest calf stands quite well without help; within 24 hours it can run with the herd. In fact, it can ill afford not to, for at any moment the herd may be pursued by a pride of lions.

So it is with many of the open-nesting birds to stay in the nest is to invite skunks, owls

nd other predators. No wonder the hen bobwhite quail fills her nest with a dozen or more eggs which hatch into chicks that follow their mother away from the nest almost immediately. Female birds which use open nests also tend to incubate for fewer days than do birds that nest in the safety of hollow trees or burrows. This reduces their exposure to enemies and more of them survive to become parents again.

Most reptiles bury their eggs and shuffle or slither away; the eggs incubate in whatever heat nature provides. But among birds—descendants from the reptile line—the warmth for incubation is nearly always supplied by one or both parents. The bird sheds some feathers from its breast; the area then fills with extra blood vessels to become a natural heating pad. This brood patch is then nestled against the eggs to warm them.

As the bird nestles, however, an egg can easily be rolled out of a shallow nest on the ground. So ground-nesting parents have evolved an "egg-rolling" instinct. To retrieve the precious egg, a goose arches its neck, presses the underside of its bill against the far side of the egg, and pulls toward the nest. Try rolling an egg with a knife held vertically, and you'll see why the goose weaves its head side-to-side—to keep the egg rolling straight.

A bird on its eggs can be a fierce defender against all comers—even its mate. As in courtship, many birds have evolved rituals to enable one parent to relinquish the nest to the other without squabbles. When a male grey heron returns to the nest, his mate stretches her neck upward and then stands, engaging him in a noisy ceremony. Boobies swap clumps of seaweed; gannets and penguins bow in solemn formality. A male turnstone offers a pebble; the female picks it up, drops it in the nest, and turns the incubation over to the male.

Some birds, such as chickens and ducklings, peep while still in the egg—and their mother calls in response as the hour for hatching nears. The feeble peeps foreshadow their emergence and a crucial phenomenon called imprinting, by which both newborn birds and mammals fix the identity of the mother indelibly in their minds.

Some animal parents have to feed their babies, others don't.

The gadwall duck (opposite) and the hippopotamus (above left) have one thing in common: their young are precocial—born with their eyes open, protective coat developed, and legs strong enough for them to follow their parents almost immediately. The young of the dormouse (top) and the purple heron (above) are altricial—born naked or nearly so and too weak to stand. Often, altricial babies are born blind. In all altricial species and in some precocial ones, parents must feed their young until they can fend for themselves.

Motherhood sometimes is a "burdensome" responsibility.

The scorpion (below) and the crayfish (opposite, middle) cart their young on their back and abdomen, respectively, protecting them from enemies. The female gray kangaroo (opposite, top) carries her "joey" in her pouch. When danger strikes, she runs. Under extreme stress, she may lighten her load by dropping the joey to the ground. Among baboons, males provide protection; females, transportation (opposite, bottom). As it grows, a baby leaves its mother more and more to play with other young.

She is, quite simply, the first moving object the infant sees. Thus a duckling follows the hen, a fawn follows the doe. And in like manner a hand-reared gosling follows the human it first saw on hatching; to the gosling, that human is mother. Although nurturing is a rare trait among reptiles, some alligator and crocodile mothers are also very attentive. Baby alligators and crocodiles chirp while still in the egg buried in a mound of debris or mud. Nearby, the mother hears them and digs up the eggs to free the newborns. Once the hatching is complete, she gently totes her little dragons to water in jaws that can crush the leg of an ox.

Some offspring find security in a fixed location—a nest or burrow. Others find it as hitchhikers in or on a parent's body. Young baboons and scorpions ride their mothers' backs; larval crayfish hang onto their mother's legs and abdomen. Bats cling to their mother's fur or nipples, even as she zips through the twilight on her hunting forays. Sea catfish hatch in the male parent's mouth and live there for weeks, dining on the plankton he inhales. While they fatten, he fasts. When they leave, he eats.

Young sea horses, too, start life with father rather than mother. Into the pouch at the base of the male's tail the female squirts her eggs. They hatch in about eight to ten days, but the babies don't leave until father muscles them out several weeks later. He may eat a few, but perhaps 150 swim off to start life on their own.

The father sea horse's pouch is merely a safe haven for the young. The pouch of a marsupial is that and much more, for only females have them and inside are the teats that nourish the tiny newborn. There the bee-size newborn kangaroo huddles for about four months. Once it leaves the pouch, mother and young develop a coordinated response to danger: she leans forward, it dives in headfirst.

One of the more comical ways in which immature animals gain security is the "caravanning" transportation system of the European shrew. The young shrews line up behind their mother, each biting the fur at the base of the tail of the shrew ahead so as not to get separated from its all-important parent. Off they go, like a child's choo-choo; pick up the mother and you pick up the whole train from engine to caboose.

In some species, unmated adults help parents with babysitting.

Among birds living in dense populations, crèches — or rearing groups — are sometimes formed to care for the young. As soon as little emperor penguins (left) can waddle, they are often found in a living corral formed by adults who have not gone off in search of food. This huddling helps protect the young from icy winds and hungry skuas and may prevent their wandering off.

The adult Canada goose above may or may not be leading a creche, for Canadas don't always form them. When they do, the number of goslings usually ranges from ten to twenty. Whether rearing a creche or only their own family of four to six goslings, parent geese keep them together so they can lead them to food and water, and warm them at night and on stormy days.

See one emperor penguin chick, we humans might say, and you've seen them all. And the emperor penguin parents might say the same. Returning to the rookery from a fishing sortie, they find the chicks crowded together by the hundreds. Out of this horde of look-alikes, the parents find their own — not by sight but by calling to it and listening for its answer. And they feed one young bird, and only one.

The parental urge to feed a chick is instinctive, but recognition of one's own must be learned by some species. The herring gull feeds any youngster of the right age and species for its first few days as a parent. But then the gull learns to recognize its own chick, and thereafter feeds no other.

Newborns that must fend for themselves often die before finding their first meal. Since nurtured young are spared that fate and need not learn to be breadwinners right away, their early energies can be poured into fulfilling other needs. Birds complete the development they could not finish in the egg. Other animals explore their surroundings and learn to avoid dangers that may lurk there.

There are many ways to feed a juvenile. Some parents lead the young to the food. Many more bring the food to the young, and some deliver it already chewed and in a semi-liquid state. When a pelican alights at a nest full of squabbling young, it is assaulted by poking and jabbing bills. The adult opens its bill wide; a chick thrusts its head as far as possible down its parent's gullet, forcing it to regurgitate partially digested fish; the chick gobbles the delicious, nutritious goo.

Herring gull chicks peck at a red spot on the parent's bill, and that triggers the parent's regurgitation response. But the blind and scrawny babies of many perching birds are incapable of such coordinated action. So nature equips some of them with a simpler reaction called the gape. The tap made by an arriving parent's feet on the rim of the nest stimulates all the little heads to shoot skyward, their outsized beaks open. So strong is the urge to stuff that gaping maw that some birds will stuff almost any similar maw as well: that of a chick of another species, of another adult bird, even of fish used to being fed by humans.

167

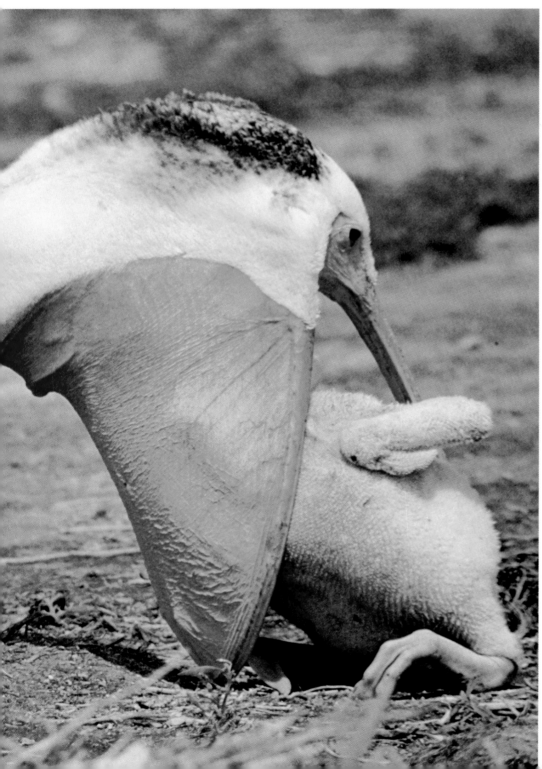

Cries of hungry offspring compel parent birds to bring food.

When parent birds see the gaping mouths of their young, they instinctively bring food. The task is exhausting—many parents lose up to 20 percent of their weight during the breeding season. It's no wonder. The blue tit (opposite, top) may make 27 trips per hour to bring insects and seeds to its nestlings. So many trips are needed because the bird can hold only small bits of food in its bill. Pelicans (left) reduce the number of trips by carrying the food in their stomachs. A youngster reaches far inside its parent's yawning maw to gorge on the partially digested, regurgitated feast.

A parent screech owl (opposite) brings mice and other small animals to its nest. The owlets swallow the mice whole. In similar fashion, common terns feed whole fish to their young (top, right). Sometimes the chick is shorter than its meal. Nevertheless, the parent stuffs the minnow down the chick's throat, leaving half of the fish dangling outside until the baby's digestive process makes room for it.

Nursing is a singularly important task of a mammal mother.

Until her babies grow teeth and eat solid foods, a mammal mother keeps them alive by nursing them. Whether a Japanese macaque (opposite), a whitetail deer (below), or an African wild dog (bottom), the essentials are the same. Most babies are fed on demand *and the milk provides nutrients for growth and, often, antibodies against diseases.*

The act of nursing yields emotional nourishment as well, for the mother's presence reassures the infant. The physical contact between parent and child also strengthens the bond that holds them together through the early lessons of life.

The strategy of bringing food to the young reaches its zenith with the mammals, for the mother suckling her progeny is herself the food supply. If her own food intake remains constant, she produces plenty of milk. And in its rich mixture of vitamins, fat, protein, sugar, and salts, the suckling of each species finds a formula exactly suited to its needs. Seal milk contains almost no sugar, but it is nearly four times as high in protein and nearly twelve times as rich in fat as the milk of domestic cows. Consequently, the shivering newborn seal quickly builds up a warm insulation of blubber. Weighing approximately 75 pounds when born, a Weddell seal in the Antarctic tips the scales at nearly 300 pounds when it is weaned, a brief eight weeks after first taking milk from its mother.

One of the newborn mammal's first tasks is to find the nipples. The primate mother makes it easy; she simply clasps her baby to her breast. The hoofed youngster instinctively searches at the angle of leg and belly; if it chooses a front leg, mother nudges it to the rear. Feline mothers usually lie with legs outstretched in a kind of corral that guides the blind kits to the spigots.

The lesson is not always so easily learned, especially for marsupials. After a gestation period of only 13 days, the embryonic opossum has tiny, clawed front legs, a mouth, and not much else—yet it must leave the womb, drag itself through a jungle of hair, somehow find the pouch entrance, wiggle in, and clamp onto a nipple before all are taken by its siblings. Though there may be as many as 17 teats, up to 21 progeny may be competing for them. But even under normal circumstances when there are enough nipples to go around, one misstep or delay spells death, for mother opossum seems oblivious to the wasp-size babies for whom she has just cleaned out the pouch.

In one regard, the marsupial newborn appears to have it easier than its mammalian cousins which have to suck and coax the milk out of breast or udder. The fetuslike kangaroo or opossum only has to latch onto the nipple. Once that is done, the nipple swells to fill the mouth and it is thought the mother pumps milk into the awaiting infant.

*Defenseless offspring depend
on their mothers for protection.*

*Young mammals need time to
learn how to feed and defend
themselves, and during that
time their mothers stand ready
to fight for them. The fierce,
700-pound female polar bear
(opposite) makes walruses—
and even the cub's own father—
leery of getting too close.*

*Kicks, not claws and fangs,
protect the giraffe calf (below).
When a predator approaches,
the calf runs underneath its
mother's belly and stands at
a right angle to her, giving her
plenty of room to strike out
with both front and hind legs.
Her kicks are so powerful
they can badly wound or
kill adult lions.*

Mammal youngsters have much else to learn
before being fully ready to cope with living
wild. A fairly large proportion of their lives
is therefore spent in juvenile dependency—
nature's classroom, in a sense. In that sheltered
status they watch and imitate their elders,
play and tussle with their siblings, and test
themselves in some of their future roles: breadwinner, communicator, defender.

But they aren't defenders yet, and defending
them is an important part of the parent's role.
Into the mammalian mother the millenniums
have implanted a ferocity in defense of her
young that is seldom matched by the male.
Indeed, since the upbringing of mammal youngsters usually requires only one parent, the
father is likely to be off by himself.

Like defending mothers of every stripe, the
sow bear's first move is to put herself between
her cub and the source of danger. Wilderness
hikers know the worst place to be is between
the two, so they wear "bear bells" whose jan-
gle heralds their approach in time for the
sow to lead her cub away.

In a den burrowed in the snow, a polar bear
gives birth to her cubs—usually two, each a
foot long and blind. There they spend the
dark winter months curled next to their
mother's warm body. If the den is threatened or
damaged, the mother may try to protect her
cubs by hollowing out a new one and carrying
them to it in her jaws. For that is the way of
the carnivores: powerful jaws that are deadly
weapons in the killing of prey are used as
gently as a gloved hand when lifting a baby.

The giraffe builds no shelter for her young,
for it is the animal's life-style to keep moving.
At birth the mother's lanky body is the newborn's downfall—literally, for she drops it
while standing up. The baby giraffe hits the
ground on its forelegs, its sharp hooves still
padded with the jellylike substance that pre-
vented their ripping open the fetal membranes
before birth. Soon the calf is able to run and it
joins other calves in play. For the rest of its life
running will be its main defense. Yet at first
the mother's body is itself a sort of shelter, a
pillared gazebo under which the calf finds
shade, security, and sustenance—until it out-
grows its overhead clearance.

By touching, mammal mothers make infants feel secure.

Whenever a bison calf feels the need, it nuzzles up to its mother for a moment of closeness (below). At about six weeks of age, a red fox kit (opposite, top left) ventures from its den. But at the first sign of danger, it retreats to the den and mother for a reassuring touch.

If close contacts between lion cubs and their mother (opposite, top right) are not renewed frequently, a stray cub may be overlooked and left to starve or be killed.

Mother chimpanzees (opposite, bottom) show few such lapses in their attentiveness. Screams, whines, shouts, and barks add to the communication between these intelligent mothers and their babies.

But for all mammals there is no substitute for the bodily contact that reinforces the bond created between them at birth—a bond that must last throughout the offspring's period of dependency.

The greater rhea, of Brazil and Argentina, also depends on running for its defense; this four-foot-tall bird long ago lost the gift of flight. But it is the male rhea that assumes the chores of parenthood. To its trampled depression in the grass, the male rhea first lures several females by dashing about, making loud, bellowing roars, and spreading his short wings. He mates with each one, then for five weeks he warms the 20 or 30 one-pound eggs they leave behind, warning away all comers, including his former harem. And when intruders menace, this bird that could flee with the speed of a horse chooses to stay and protect his young.

We humans are tempted to credit the male rhea with noble qualities: bravery, devotion, self-sacrifice. And we are even more tempted to read our own emotions into the touchings of mammal mother and babe. The bison calf nuzzling its mother, the lioness nudging her cubs, all seem to be expressing affection. More likely, though, they are simply reinforcing the bonds that keep them together. To the young bison, the feel of its hulking mother probably imparts security. And to the mother bison, the feel of her calf close by probably reassures her that all is well.

It is in the primates that we observe bonds that seem most clearly akin to our own. Surely more than recognition and reassurance pass between the mother chimpanzee and the boisterous babe she dandles.

Surprisingly, many animal societies that form rigid dominance structures show great leniency toward obstreperous youngsters, until it's time for them to become members of the adult social order. A lion cub is permitted to chomp his majesty's tail with near impunity. Young howler monkeys are allowed to brawl and clamor until adult nerves begin to fray. Wolf pups romp with their elders without regard for rank; later they will have to snarl and snap their way into the hierarchy they so blithely ignored.

Amid the sea's brutalities, it is refreshing to come upon a show of tenderness by a mother toward her young. Not surprisingly, the best examples are mammals; in no other order are the bonds between parents and offspring so

consistently strong. Nature knows no relationship so intimate or profound as that of the babe formed within its mother's body and sustained thereafter by its mother's milk.

Life for the sea otter begins and ends in the sea. A strong bond between mother and pup quickly develops at birth. For months they are practically inseparable. Clinging to her back or clasping to her breast, the pup rides along as she goes about her daily routine. Once it is weaned, it can count on a share of the food she gathers. And if it ventures too far from her side, it need only call her with its little yelps and trills, and she will come.

The calf of the manatee of Florida shallows seems to be almost coddled with attentive care. Born in the water, the calf weighs in at 60 pounds. But it seems light and buoyant as its father—a seven-foot submarine of some 400 pounds—cradles it in his flippers or passes it to its mother for nursing. The calf sucks at her leathery nipples while the father leaves to eat lush vegetation in slow-moving waters nearby. Then he returns and cradles the young sea cow while its mother forages. Between the efforts of the two of them, the babe is cradled almost constantly. And nearly two years later it will still be nursing. Life plods slowly for these gentle vegetarians.

Life among the carnivores proceeds at a more hectic pace, for the growing young bread winners have many skills to sharpen and many lessons to learn in the mastery of their hunting craft. Littermates who once suckled peacefully side by side begin to scuffle and snarl in games that mimic the hunt and thus hone their readiness for it. Stalking, pouncing, running chasing, biting—all are practiced now in play as they will be used later in earnest.

Sooner or later the skills learned in play are used in survival and to acquire status Among social animals such as wolves and foxes maturing youngsters must fight to reach their rung on the dominance ladder. And among loners such as the bears, a developing aversion to companionship turns sibling on sibling in flurries of fang and claw and growl.

Amid the hubbub, the all-important learning process goes on. Its most important factor i the firm bond long since established betwee

Playful young practice skills needed to be successful adults.

The red fox kits playing a game of chase and escape (top) are getting more than exercise from their sport. They are learning how to time leaps at moving prey and how to maneuver to keep from becoming the prey of larger meat-eaters. Though all mammals play when young, the

types of games reflect the needs of the species playing them. The lion cub harmlessly biting the neck of its sibling (above) is learning the fine points of feline fighting and killing techniques. The nearly grown Alaskan brown bears (opposite) are practicing the threat displays that, in adult life, may deter injurious battles.

176

The mother-offspring bond decreases as the young mature.

The Alaskan brown bear cubs (below) have joined adults feeding at a river full of migrating salmon. As the bond between the cubs and their mother weakens (partly due to her readiness to mate again), *the cubs may leave her to gather around another adult female to nurse or observe her as she fishes. Shared rearing of cubs is also very common among lions. The lioness teaching cubs to locate water (opposite, bottom) probably has her own litter plus one or two playmates.*

In sharp contrast to the carnivorous bear and lion cubs, *which take about two years to gain independence, the herbivorous—grass-eating—llama (opposite, top) stops nursing after six to twelve weeks. The survival skills required by this formerly wild grazer are less sophisticated than those of lions and bears, and it needs far less time to master them.*

mother and young. Whither she goes, this bond ensures that they will follow in a rush. From her come rewards for proper actions and punishments for foul-ups; many an errant cub has been sent sprawling by a swipe of the parental paw. In truth, she is a walking textbook, a living audiovisual aid in the instruction of the juveniles.

Mother bear leads her cubs to the river's edge where they may observe her snatching salmon. She wades in, watches, makes a grab—and suddenly she has food for them. The cubs bubble with excitement, and when next she enters the water they will watch more closely to see what she does. Later they will begin to imitate, and eventually they will succeed.

Is she teacher, or just example? Little proof exists of a conscious attempt by an animal

mother to instruct her young. But as she hunts, forages, builds, communicates, and even mates and gives birth, the youngsters watch and imitate. She may not will them to imitate, but the effect is the same as if she does.

Creatures like reptiles that are guided entirely by inborn behavior patterns—instinct—need neither lessons in nor examples of how to live. Their only need for parental care is for protection while still in the egg. Little or no bond develops in their short period of dependency; in fact, many experience no such period and never see their parents at all.

It is the animals further along the evolutionary trail that exhibit more complex patterns of behavior. And since much of their behavior is learned, for youngsters that means a longer period of dependency on parents.

Mother lion leads her cubs to water. Until she does, the suckling infants know nothing of water; they get their moisture from mother's milk. So the first trek to the river or the water-hole holds many lessons: what open water is, how to find it, how to avoid any dangers along the way, perhaps even a brief hunting lesson if small prey happen by. Any exposure to a new situation can be a valuable learning experience, and the young need many of them before taking off on their own.

For the hunted as for the hunter, there is much to be learned. But the children of the hunted have greater need for protection and security because they are prime targets in the predator's eye. So for them, more of the long infancy is spent learning how to defend themselves than in how to find food.

There are times when even the most devoted mother must leave her babes alone. Mammal mothers have different ways of preparing their young for the hazards involved. A black bear trains her cubs to climb a tree whenever danger threatens and not to come down until she sounds the "all clear." This life-saving drill stands the young bears in good stead long after the mother has abandoned them. They continue to climb trees until they discover they can cope with danger as well as she did.

In countless ways, the role of the parent determines the way of the child. The parent gives it life and equips it with all it needs for the journey to maturity and parenthood of its own. What the child owes to its parents, it then repays to its children—the secrets of survival are passed on, the species carries on.

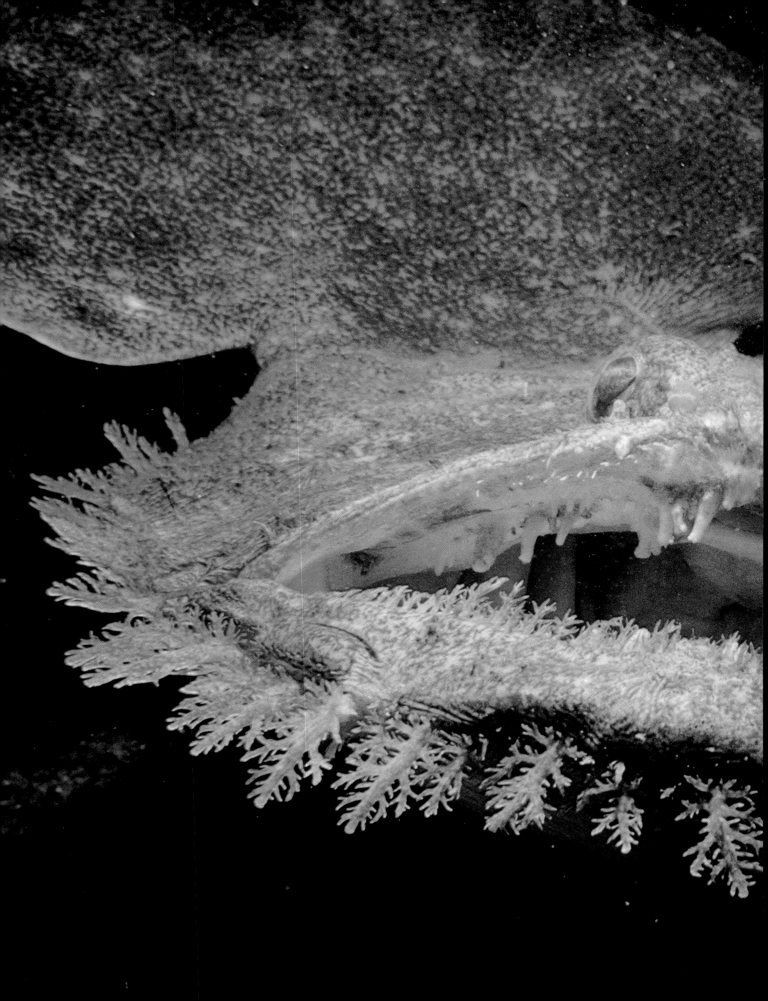

Pushing the Limits

The ever changing sea
and land force animals to
find new-and sometimes
bizarre-ways to survive.

The Innovators

Four miles above sea level, a mountain climber labors toward the snowy summit of Mount Everest. Insulation pads his body against the knifing winds; tanks of oxygen sustain him in air too thin to breathe. Surely the planet Earth knows few environments as harsh as this. Then he glances down—and there in a crevice of the barren rock squats a jumping spider, its rows of eyes filled with this strange apparition from the world below.

Protégé of the wind, the spider waits to pounce on insects blown to its aerie from the life-filled lowlands. In its lonely world, others also wait upon the wind; springtails no bigger than a speck of pepper feast on an air-mailed fallout of pollen. Look in your bootprint, climber; you may see them jumping about, launched by a tiny appendage as supple as a buggy whip. And look too at the "barren" rocks around you, for there cling the lichens. These primitive plants may hold their own for weeks and months and actually grow perhaps only two days a year.

A harsh environment? There is no such thing. A salamander wiggling among wet leaves would quickly shrivel on hot dry sands, yet the horned toad would have its desert no other way. Only the artificially equipped mountaineer sees the top of Everest as a hostile world. Strip him of his padding or his oxygen bottle, and the jumping spider would soon see this visitor flunk the stern tests that the spider has managed to pass.

In a changing world, the tests are constantly changing. Willows once flourished in the Sahara. Fossil seashells sprinkle the rangelands of Nevada. New food sources appear as old ones dwindle; new enemies arise against whom the old defenses no longer work. Life on a changing planet must keep pace by making changes of its own, discontinuing outmoded species, testing new models, perfecting inventions that seem to work well—the eye, the heart, the brain, the wing.

Each creature handles the day-to-day challenges to its existence by assuming different roles and performing the skills required to fulfill them. But for the survival of its species, the most important role may well be that of the innovator, the individual that does an old thing in a new way. These are the rule benders, the animal oddities that eat the inedible, do the implausible, inhabit the uninhabitable. In strange and wondrous ways they probe the very frontiers of life.

Life: in today's world, that short word embraces about a million and a half known kinds of animals and half a million kinds of plants, plus who knows how many more to be discovered. Yet the circling eons have already weeded out more millions of species than we can accurately estimate—including 800-pound beavers, ground-shaking tyrannosaurs, ancestral horses no bigger than a fox. In the laboratories of time, most of the experiments begun to date have been concluded. Only a fraction of the species tested are alive today.

And who will inherit the earth tomorrow? Perhaps we glimpse some answers in the innovators of today. Even their names can sound alien to ears accustomed to the familiar ring of bear and bee and bluegill: addax and fennec, two mammals that can survive without drinking water; lemur and gecko, a mammal and a reptile that ride the air without wings; axolotl, a fishlike amphibian with feathery gills on the outside of its neck; anableps, a tiny fish whose bifocal eyes enable it to search for insects above the water while watching for predators below; capybara, a rodent of some 100 pounds that whistles, grazes, and hides underwater; fulmar, an arctic bird that won't leave its nest even if 50-knot winds bury it in snow and freeze its eyes shut.

Before we label them oddballs, we humans should remember that we are the most indefatigable innovators of all. As Bedouins we trudge the searing sands; as Eskimos we wrest a living from snowdrift and icepack. As breadwinners we tinker with evolution to produce plants and animals that didn't exist before. As communicators we babble in more than 3,000 different tongues. Yet if we had to compete with animals on their terms, they would outpoint us again and again.

We cannot breathe water, and neither can the fishing spider—but it can dive with its own bubbles of air and tap a food supply that is off-limits to most of its own kind.

Non-swimming spider outdistances fish in underwater chase.

Unlike the web-spinners, the vagabond fishing spider builds itself a raft of dried leaves, debris, and silk threads and sails away Huck Finn style, usually after insects but occasionally after fish three or four times its size. The 3/4-inch spider appears to attract fish by stroking the water with its legs. When its sensory hairs detect prey, the fisher uses its rare ability to skate across pond or stream or dive in pursuit.

Air bubbles trapped on its lower abdomen allow it to stay underwater for a half hour or more. Ambushing minnows or tadpoles, this non-swimmer clings to an aquatic plant until its victim comes within striking distance. After seizing prey, the fisher subdues it with poison, then injects digestive enzymes. To avoid diluting the enzymes, the spider bobs up on its air "pontoon" and lugs the fish out of water onto its raft or onto the streambank where it feeds on its catch.

Previous page: Goosefish—Atlantic floor off Maine.

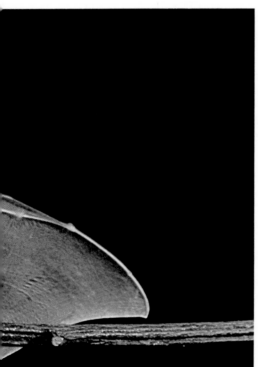

In loose, sandy soil, a man might be able to dig a tunnel with his bare hands, but a trapdoor spider digs a burrow eight times longer than its own body, lines it with silk, corks it with a lid held on by a silken hinge—and pops out like a jack-in-the-box to grab a passing meal. When danger threatens, the spider sinks its fangs into the underside of the lid and holds the door shut from within. During the molting period, the armorless arachnid sews the trapdoor firmly shut with threads as a safety measure until a new protective coat hardens over its soft body.

To devise a new way, every innovator has to undergo change. By its capacity to change, the bat has developed a new way to be a mammal: it flies. Change has given the chameleon a tongue up to ten inches long; in some species this insect-catcher is longer than the animal's entire body.

Shrew and whale stand today at opposite ends of the mammalian parade, one the smallest, the other the largest—in fact the largest creature of all time. Yet into the blue whale's cavernous mouth the great beast takes nothing larger than a shrimp. Why then does it remain so huge? Swimming slowly through the sea's vast clouds of krill and other plankton and seining out the little creatures with its baleen works so well, the whale really has no need to change its size. Streamlining for speed would be superfluous, because there is no one to run from. Its bulk alone is an effective deterrent to most attackers. Certainly attempts to return to life on land would fail, for bones strong enough to support the whale's 135 tons would be too large to let it move. But in the sea the gentle leviathan is almost weightless. Gliding serenely along, it has reached a near-perfect accommodation to its natural environment.

While the shrew is the smallest of mammals, it is also one of the fiercest breadwinners, attacking even its own kind in a lifelong marathon of gluttony. Its size is its goad, for the smaller a mammal is, the greater its surface area in relation to its weight and thus the greater its heat loss through the skin. We humans need to eat about a hundredth of our weight in a day. But a mouse must eat about a fourth—and a shrew must eat nearly nonstop.

Novel breadwinners exploit nature's endless bag of tricks.

Physical prowess, adaptability, or intricate construction skills give some creatures an edge when seeking or storing food.

To catch insects, the Jackson's chameleon (opposite, top) coordinates telescopic eyes with a darting sticky tongue. On contact the tongue contracts and the lizard gulps the insect.

When the keen eyes of the loggerhead shrike (above) spot insects, lizards, mice, or small birds, the hunter's sharp bill zeros in, often killing more than it can eat. The shrike habitually stores the excess on thorns, or, adapting to the new, it hooks its prize inventively on barbed wire.

The fringed-lipped bat's sensitive ears capture echoes of its own voice to guide it to the tree frog (bottom, right). Time interval and angle of sound pinpoints distance plus exact location of its victim.

The trapdoor spider plays a waiting game, hiding in his silk-lined tunnel under a trapdoor (opposite, bottom). When an insect passes by, the spider pounces from its "hunting blind" and pulls in the prey.

It may be that mammals can get no smaller; it appears they simply could not eat enough to stay warm.

By contrast, the heaviest insect may be the goliath beetle, a four-inch giant weighing in at about three ounces. Apparently insects can get no larger, for they breathe not through lungs or gills but through ducts that bring air directly to the tissues—and that works only when the ducts are small and simple. Those that wear exoskeletons and shed them periodically must stay small lest they be squashed by their own weight before the new shell can harden. And those that prance on spindly legs would prance no more if they grew too heavy. Even for the innovators, the rules bend just so much.

Yet within the rules there is latitude aplenty. As breadwinner, the puffer fish keeps to its normal size; some grow to perhaps three feet long. But as defender, the threatened puffer swells to a ludicrous beach ball perhaps twice its normal bulk—a bluffer's way of intimidating a predator. The porcupine fish and the sea urchin have added an extra innovation, a stem-to-stern armor of spines. A body bristling with spines serves well the defender who is ill equipped to fight or flee.

Camouflage is a far gentler defense, and innovations make it an endlessly fascinating one. What could be more delightful than the disguise of the larva of the geometer moth, better known as the inchworm or measuring worm. Parked on a blossom, it decorates itself with bits of petals so it can graze undetected by birds. The green lacewing larva can suck a woolly alder aphid dry in minutes, if it can only elude the ants guarding the aphid herd. So it hastily grabs the first aphid it can, yanks off the helpless insect's tufts of fuzz, and secures the wool to special hooks on its own back. In 20 minutes it can start dining in disguise on the aphids, a tiny wolf in sheep's clothing under the very jaws of the unsuspecting guardian ants.

Nature's innovators have often beaten us inventive humans to some of our proudest discoveries—usually by many thousands of years. How astonishing that a simple fish called the electric ray can kill an enemy or stun a man

Quick-change artists mask identity with cunning disguises.

Camouflage and alarming shapes help some animals escape detection. The electric ray (opposite, top) hides by slipping partially under the sand, flattening its wings to eliminate shadow. The unfortunate predator that uncovers his guise may receive a 220-volt shock!

Also trying to blend with its surroundings to conceal itself from hungry birds, the geometer moth larva, or measuring worm (the loop in blossom, opposite, center) festoons itself with petals and masquerades as part of the flower.

The echidna, or spiny anteater (opposite, bottom) and the puffer (below) try another defense stratagem: changing shape. For self-protection the echidna rolls up partially with only naked snout and paws visible; serious danger causes it to burrow quickly into the earth and roll into a tight spiny clump. The puffer (lower right) inflates itself into a hard, spiky ball (upper left) so its predators will think it's too large to swallow.

with a jolt of electricity. How humbling to find fish that can set up an electric field around themselves and detect prey or obstacles by the distortions in it. And how frustrating to watch for thousands of years as birds, insects, and even a few fish and fellow mammals rode the winds until we finally figured out how to get off the ground.

Mammals such as the flying squirrel don't really fly; they simply stretch out flaps of skin and glide. But as travelers these innovators cover impressive distances. Australia's greater glider possum can soar 120 yards. Its smaller cousin the sugar glider stays aloft long enough to catch moths in midair.

Flying fish don't fly either. But they seem to as they explode out of the water at speeds of up to 35 miles an hour and glide as far as 150 yards and more on huge, trembling pectoral fins. At best, the fish's "flight" lasts about 14 seconds. A fish can live out of water longer than that—but if kept out too long, it dies of suffocation as its featherlike gills collapse. By ingenious innovations, some fish push the time limit to amazing lengths. The mudskippers of Asia slog about in the ooze for several minutes at a time, holding water in their gills with special gill covers, rather like a diver holding his breath. Eels can travel short distances through grass dampened by the chill of night, breathing through their moist skins as they

wiggle along. Asia's climbing perch breathes with special pouches above its true gills; such "climbing fish" seen in trees probably were carried there by birds. But perhaps no fish out of water can outlast the lungfish species of Africa and South America. In dry spells they burrow into the mud and breathe with their lungs until the waters finally return; one was kept this way unharmed for several years.

Such fish must have compelling reasons to come ashore, for swimming is one of the most effortless ways to travel. Flying costs an animal much more of its energy, and running costs more than both. But the costliest yet measured by man is also one of the slowest: the lethargic creep of the gastropod. The slug not only must generate a constant flow of mucus; it must also overcome the slime's stickiness in pushing itself along by waves of muscular contractions in its foot. Ounce for ounce, this traveler pays 12 times the fare in energy that a running mammal does. There are a few dividends, though, for the foul-tasting slime does repel many predators and helps the slug to track down a mate.

Innovators often pay high prices for their ingenuity. To master the sea, the emperor penguin lost the gift of flight but became the diving champion of birddom, routinely plunging to depths of more than 800 feet to catch the fish it lives on. Heavy bones help it de-

Squirrels and fish take to the air; catfish trek across land.

Certain species break out of their earth- or waterbound lives to gain unusual freedom in their modes of travel. The flying squirrel (opposite) soars from tree to tree by using "hang glider" flaps of skin between the fore- and hindlegs. To avoid branches or check speed, it tenses or relaxes the flaps and slants its tail.

Flying fish (below) flee dolphinfish by propelling themselves out of the water. Gaining impetus by vigorous swimming, they break the surface and sail 150 feet or more before retouching the water, then regain momentum with a tail-fin sculling motion on the water to become airborne again.

When a pond dries up, walking catfish (top) wiggle on spiny fins in search of water. With special sacs surrounded by blood vessels in its head, the catfish gulps enough air to last several hours.

From bubble dome to condo, creative builders fill their needs.

Some animals weave homes to suit their life-style. Sociable weaver-birds (opposite) and harvest mice (top right) weave their homes of grass and straw. The gregarious birds first build a massive nest, then by nipping off straws, the 20 to 30 mated pairs hollow out apartments.

The harvest mouse weaves her thatched home high and dry amid tall stalks (top right). The tiny rodent keeps a neat house until the rambunctious young push their way out, tearing it apart, so mother rebuilds for each new brood.

Caddis fly larvae (top left) and water spiders (above) make their underwater homes of silk. The caddis fly larva spins a sheath to cover its soft abdomen. It disguises this mobile home with pebbles, twigs, and shells.

The spider spins a sheet and attaches it to plants with threads. To fill its need for oxygen, it traps air bubbles at the surface and carries them between legs and body to inflate the sheet into a "diver's bell." It catches prey from its nest or by rising to the surface, returning to the bell to feed.

scend, and a huge lung capacity enables it to remain underwater for 18 minutes at a time—another record for a bird. With its flipperlike wings it still "flies" through the water at about ten miles an hour, a poem of grace and efficiency. And once ashore, it and the others of its penguin kin are the only birds in the world that can walk, or shuffle, upright—that is, with a vertical backbone.

The common swift too walks only with difficulty; so it compensates by spending nearly all its days in the air, swooping after insects as it feeds, endlessly circling until the sky darkens into night. If it weren't for the need to nest, it would hardly touch down at all.

Nest it must—and here too the swifts are innovators, for their sticky saliva serves them as nest cement. The palm swift of Africa makes the absolute most of that asset; it glues its one or two eggs directly to the vertical surface of a hanging palm frond, clings upright to the frond as it incubates them—and finally hatches out chicks that must cling to the same precarious perch on the side of the frond until at last they can fly away.

Birds that push the limits in the other direction find that too much nest can be precarious too. Communal nesters are somewhat rare in the bird world, yet their overzealous architecture has brought many a sturdy branch crashing to the ground. Like the sociable weaver of Africa, the groove-billed ani of the American tropics doesn't seem to know when enough is enough. To their communal nest, several pairs of anis keep on adding leaves and sticks even after laying has begun. Some eggs simply get buried under the new carpeting; instead of hatching and becoming part of the flock, they become part of the nest.

A neat little sphere of soft grasses suspended over a meadow on sturdy stems—now that's a proper bird's nest, a fitting cradle for a clutch of eggs. Then what's that harvest mouse doing inside? Innovating. This gifted little builder weaves a nest to rival any bird's, a rare trait among mammals but a valuable one in habitats where burrows are frequently flooded.

The life-style of the loners also produces some strange innovations. The nearly motionless three-toed sloth is hard to distinguish from

Cooperators as well as loners lead non-conformist lives.

Although sociable in their growing years, adult orangutans (opposite) and giant pandas (above) are true hermits; they socialize only at breeding time and with their young. To warn other males away, orangs resonate a call a mile through the jungle. Accidental meetings result in violent branch shaking until one male retreats.

Pandas take no time for togetherness. They spend 10 to 12 hours a day eating their way through bamboo thickets trying to extract enough food value from this high-fiber plant.

Sea anemones and clown fish (left) enjoy a symbiotic relationship that is a far cry from the predator role that the anemone plays with the blenny (see page 20—Ed.). Protective mucus on its skin allows a clown fish to take shelter among the anemone's stinging tentacles. The clown fish has been seen repaying its host by feeding it or driving off its predators.

Ants and aphids similarly barter services. In exchange for the aphids' honeydew, the ants shelter the aphids from their predators and eat their predators' eggs. To ensure more honeydew, they stroke the aphids and also herd them to more nutritious plant tips.

the foliage of South American jungles because of a green patina of algae that grows on its fur. Here is a stone that rolls through life very, very slowly. A meal of foliage may take a week to digest, and the animal can wait at least that long between descents from the treetops to urinate and defecate. On the ground it drags itself laboriously along with its stout, curving claws; in the branches it hangs by them belly up, swiveling its head through 270° of turn as it sleepily surveys its domain. A lovestruck sloth does not even leave the trees to mate. Taking his own sweet time, he need only listen for a come-hither whistle from an interested female and pursue the "follow me" scent messages she has left on limbs and vines.

Another jungle dweller who communicates with its nose is the proboscis monkey, but this sociable Cyrano uses it to produce sound. The male's loud nasal honks, reinforced by his ballooning snout, ricochet through the rain forests of Borneo. But for a truly exotic, hard-to-crack code, the way one honeybee tells another where the flowers are is a marvel of change and adaptation.

On returning from the field, a forager bee reports to its hive mates the direction and distance to a new source of pollen and nectar by performing intricate dances. A round dance signifies a near-at-hand source. A waggle dance refers to more distant blossoms and consists of figure-eight patterns executed in a frenzied way. The dancer circles both to the left and to the right. On the straight intersection of the two circles, the bee flies on a specific angle to communicate the exact number of degrees in the angle between the food-rich field and the sun.

The speed of the dance also communicates the distance of the source from the hive: the faster the tempo, the closer the field. The number of abdominal waggles, or violent shakes, the bee makes on the straight run and the duration of its buzzing may also indicate distance. All the while, the forager's attentive audience gathers round with their antennae pressed close to pick up her messages.

No one likes to acknowledge the awful act of cannibalism even though it does happen in nature. But the act is made even more alien to us when coupled with the act of mating. Yet

Courtship habits range from gentle to seductive to dangerous.

Males of different species use varied mating approaches. The fiddler crab (opposite, bottom) says it with his "fiddle," a large claw that he waves or taps to get a female's attention.

As a measure of his ardor, the ostrich's slender neck distends and blushes a deep pink (below). He offers himself to his chosen one by strutting, swaying, spinning and waving his spread wings, finally crouching before her.

The praying mantis (right) is justifiably much more conservative in his approach. With a healthy fear of the larger, carnivorous female, he takes about five minutes to move an inch, until he is close enough to clasp her—before she can grab him with her forelegs and bite off his head as happened here.

the female praying mantis often turns and seizes her mate, sometimes devouring his head even as his nether end continues to fertilize her eggs. Actually his life cycle is nearing its end anyway, as will hers shortly after her foamy egg sac is extruded onto a twig. At least his body serves to nourish hers and the eggs in which they pass the gift of life on to the next generation.

The female horseshoe crab takes a very different approach to her suitors. One, two, even five may hook their shells onto hers at mating time. Slowly she drags her entourage up onto the beach, digs a hole, and lays eggs for any and all to fertilize. We can hardly call this a new idea, for it has surely been the way of this ancient creature—not a crab at all but a cousin of the spider—for all of its 200 million years. Yet old ideas can strike us as novel and the view back into time as strange and wondrous as a glimpse ahead.

How long has the male fiddler crab brandished its left claw to attract a mate? Long enough for the claw to evolve into an appendage useless for eating; long enough for each species of fiddler to work out its own unique semaphore as insurance against cross-breeding. How long have bighorn rams won their ewes by butting each other hour upon hour, blow by resounding blow? Long enough to develop a double-walled skull, buttressed with bone and padded with heavy hide. How long have octopuses and cuttlefish enjoyed some of the sharpest eyes of all the invertebrates? Long enough to evolve courtship strategies that depend on this keenest sense. The courting cuttlefish displays a pattern of zebra stripes; some octopus males go them one better by courting a female with a pattern of stripes that flash from horizontal to vertical, then fertilizing her before the fascination wears thin.

Gray phalaropes seem to have the standard courtship equation backwards, for it is the females that woo the males—and the males that hatch the eggs. Arrayed in brighter plumage, the females squabble over the drab little males, courting them with raucous cries and buoyant aerial displays. A harried male may drive many of them away before finally signaling his choice with a touch of his beak on her

*Innovative parents find unique
ways to bring young into world.*

*The female sea horse deposits
as many as 600 eggs in
a brood pouch on the belly
of the father (opposite). In
it he fertilizes and feeds the
eggs from his "pregnancy"*

blood until muscles rhythmically expel each "colt."

*The male Mallee fowl (bot-
tom) also incubates the eggs.
First both parents build a
large mound nest of fermenting
compost covered with sandy soil.
In a crater in the middle the
female lays eggs at four- to
seven-day intervals from about
August to February. Each time
she lays an egg, the male
opens the mound by kicking*

*away the soil. To keep the
temperature at 90° to 95° F.
he performs a feat of thermal
engineering, opening and
closing the nest with his feet
to adjust the temperature.*

*The female Nile crocodile (be-
low) also buries her eggs in
a compost nest of sand. At hatching time she cracks the eggs
open gently with her teeth and
carries the young to water.*

breast. She lets him choose the nest site, for he
alone will tend it. And while he sits on her
clutch of eggs, she flutters off to win another
mate.

Good parents are rare among fish; "pregnant"
males are rare among any order of creatures.
So the male sea horse with its pouch full of
eggs doubly earns a place among the innovators; just being a sea horse is oddity enough.
But among the myriad ways to be a father,
perhaps even more unusual than the sea
horse—certainly more demanding—is that of
the mallee fowl of the dry southern Australian
scrublands.

In April or May the male mallee fowl and his
mate begin scratching with their big, powerful
feet. When they have dug a hole three feet
deep and up to fifteen feet in diameter, they
fill it with leaves and twigs. After rains patter
down and dampen the vegetable matter, the
birds rush to cover it with sand and dirt. Then
the vegetable matter begins to rot and that
generates heat. In September when the male
judges the temperature inside the compost nest
to be right (90° to 95° F.), the female begins
to lay eggs in an egg chamber in the heart of
the mound which he opens up for her. For
about six months she lays an average of one
large egg a week, each weighing about one-
tenth of her body weight.

Now the long grind of incubation begins for
the male. He fusses over the huge mound,
poking his beak in to test the temperature with
his tongue. Sun too hot? He adds sand for a
heat shield. Fermentation slowing down? He
takes sand away and lets the sun warm things
up. Test. Add. Test again. Take away.

Actually digging from seven to thirteen
hours a day, the harassed male doesn't notice
when the eggs begin to hatch after seven
weeks' incubation. Indeed the female pays no
attention to them either. Each chick digs its
way out of the mound and takes off, completely
feathered and self-sufficient. At year's end
father takes a well-earned vacation. About a
month later they begin all over again.

As individual animals have pressed the limits
of their abilities, their feats fill nature's record books with some pretty impressive innovations. But the grand champion, the oddity

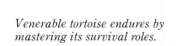

The anatomy of the giant tortoise has served it so well that it has changed little from its ancestors of 200 million years ago.

As a breadwinner it can go without food and water for weeks, and on arid land some species have evolved a longer neck and forelegs and a raised shell to allow it to browse on higher vegetation.

As a defender, it has depended upon its shell. Made of bone with a hornlike covering, the growing shell guards the soft inner parts and partially retractable head and limbs. The tortoise does not molt, so has no vulnerable periods. When damaged, the shell regenerates.

As a traveler, the tortoise uses stumpy clawed feet and good equilibrium to move over rough Galapagos terrain from its lowland habitat to highland feeding grounds.

As a courtship strategist, the tortoise mates internally, and the female can store sperm for years, so can reproduce in the absence of males.

As a parent, the female buries her eggs, then erases surface clues to foil nest raiders.

Only 11 of the original 15 Galapagos races are known to survive; but with its endurance and with our commitment to protect endangered species, we can hope that the giant tortoise will be at home on earth for eons to come.

among oddities, the innovators' best-in-show is the duck-billed platypus. Here is a warm-blooded animal that lays eggs, yet nurses its hatchlings with milk oozed from teatless mammae. It walks with the sprawl of a reptile and swims with a tail like a beaver's. It hunts for worms and crustaceans with a sensitive bill that seems to belong on a duck, eating about half its own weight in a day. Males even have fanglike poison spurs on their hind legs. It is as if nature toyed with the leftovers of evolution and suddenly they shuffled off to find a mate.

Scientists at first thought the creature was a hoax; when given one to examine, they yanked at the bill they thought must be held on with glue or stitching. Once accepted, the animal was assigned to taxonomy's pigeonhole for egg-laying mammals—the monotremes.

In the platypus we may catch a glimpse of an ancient link between the reptilians and the mammalians. If so, the animal is a glance backward, a look over our shoulders into the mists from which we emerged. Another is the coelacanth, an incredibly ancient fish species in whose limblike fins we can glimpse the rude beginnings of the human arm. More familiar to us is the giant tortoise of the Galapagos Islands, survivor supreme, in whose mighty fortress the slow heartbeats often thud on into a second century.

Once we thought all life drew its energy from the sun. Now we discover that geothermal energy may also be a life-giving source. Tubeworms five feet long have recently been found at vents on the floor of the sea, synthesizing their own food with hydrogen sulfide, formed from the union of scalding salt water and chemicals in the rocks. So we have to wonder what other alien ecosystems yet await discovery at the frontiers of our knowledge.

We cannot know what new life forms will inherit the earth; we cannot know what sort of earth they will inherit. But we humans are the inheritors of the present, and as such we share the challenges faced by our fellow passengers on the ark of earth. Many species are pushing the limits of their capacity to adjust to our pressures on their lives. What role will we play in adjusting to their needs?

Index

Picture Credits

Page 1: Murres and kittiwakes on seacliff, Cape Lisburne, Alaska, John Stern/Animals Animals. 2-3: Caribou crossing Kobuk River, Alaska, David C. Fritts/Animals Animals. 4-5: Red squirrel, Martin W. Grosnick/Bruce Coleman, Inc.

Eating Without Being Eaten
Pages 8-9: Lion chasing Thomson's gazelle, Baron Hugo van Lawick.

The Breadwinners
Page 11: Peregrine falcon, David F. Boehlke. 12: Bison, Lowell Georgia. 13: Pika, Stephen J. Krasemann/DRK Photo. Pronghorns, Charles G. Summers, Jr. 14: Honeybee, C. Allan Morgan. 15: Cynthia silk moth caterpillar, Andrew Skolnick. Eastern cottontail rabbit, Karl H. Maslowski. Giraffe, M. Philip Kahl/Photo Researchers. 16: Cheetah, E.R. Degginger. 16-17: Cheetah stalking Thomson's gazelles, Tex Fuller/Animals Animals. 18-19: Golden orb weaver with grasshopper, Donald R. Specker/Animals Animals. 19: Golden orb weaver with dragonfly, Wendell Metzen. 20: Sea anemone closing on fish, Heather Angel. Sea anemone closed, Jeff Foott. Octopus, Tom Stack/Tom Stack & Associates. 21: Man of war, Runk/Schoenberger from Grant Heilman Photography. 22-23: Leopard seal and Adelie penguins, Francisco Erize. Seal capturing penguin, Francisco Erize. 23: Killer whales and seal, Francisco Erize. 24-25: Bobcat chasing snowshoe hare, Stouffer Productions/Animals Animals. 26: Chimpanzee, Warren and Genny Garst/Tom Stack & Associates. 27: Egyptian vultures, Norman Myers/Bruce Coleman, Inc. Sea otter, Jeff Foott. Galapagos woodpecker finch, Miguel Castro from National Audubon Society Collection/Photo Researchers. 28: Peringuey's viper crossing sand, Karl H. Switak. 28-29: Peringuey's viper covered with sand, Karl H. Switak. 30: Brown pelicans, James H. Carmichael/Bruce Coleman, Inc. 31: Brown pelican pursuing fish underwater, Bill Curtsinger from National Audubon Society Collection/P.R. 32: Dung beetle, E.S. Ross. Squirrel, John J. Dommers.

33: Vultures and hyenas, Norman Myers/Bruce Coleman, Inc.

The Defenders
Page 35: Black rhino, Leonard Lee Rue III. 36: Colorado River toad, Zig Leszczynski/Animals Animals. Sphinx moth, Leonard Lee Rue III. 37: Stonefish, Zig Leszczynski/Animals Animals. Long-tailed weasel, C. Summers/Tom Stack & Associates. Willow ptarmigan, Leonard Lee Rue III/Animals Animals. 38: Acraea moth, E.S. Ross. Opossum, Allan Roberts. 39: Hognose snake, Alan Blank/Bruce Coleman, Inc. 40: Arrow-poison frogs, Zig Leszczynski/Animals Animals. 41: Stinkbug, C.W. Perkins/Animals Animals. Musk turtle, Zig Leszczynski/Animals Animals. Garden tiger moth, Jane Burton/Bruce Coleman, Inc. Coyote and striped skunk, Jen and Des Bartlett/Bruce Coleman, Inc. 42: Bumblebee, Carroll W. Perkins. Robber fly, Carroll W. Perkins. Monarch butterfly, Oxford Scientific Films/Animals Animals. Viceroy butterfly, Stan Schroeder/Animals Animals. 43: Eastern coral snake, Breck P. Kent/Animals Animals. Scarlet kingsnake, Zig Leszczynski/Animals Animals. Ant-mimicking spider, E.S. Ross. Ant, Raymond A. Mendez/Animals Animals. 44: Badger, Tom Brakefield. 44-45: Armadillo, Leonard Lee Rue III. 45: Coachella Valley fringe-toed lizard, Zig Leszczynski/Animals Animals. 46: Impalas, M. Philip Kahl/Verda International Photos. Springboks, Sullivan and Rogers/Bruce Coleman, Inc. 47: Ostrich, M. Amin/Bruce Coleman, Inc. 48: Alligator, Stan Osolinski. Green tropical tree snail, James H. Carmichael, Jr. Tan tropical tree snail, James H. Carmichael, Jr. 49: Pangolin, Hugo van Lawick/Bruce Coleman, Inc. Porcupine, G.E. Robbins/Tom Stack & Associates. Box turtle, Breck P. Kent/Animals Animals. 50: Sea star, Tom Stack/Tom Stack & Associates. 50-51: Green anole, James H. Robinson. 52-53: Lioness attacking zebra, George B. Schaller. 53: Mongoose attacking cobra, Norman Myers/Bruce Coleman, Inc.

Seeking a Place
Pages 54-55: Olive ridley turtles, David Hughes/Bruce Coleman, Inc.

The Travelers
Page 57: Snow geese, Robert P. Carr. 58: Desert locusts, Gianni Tortoli from National Audubon Society Collection/P.R. 59: Ladybirds, Keith Tucker. Spiders, John L. Tveten. 60: Monarch butterflies, Jeff Foott. Monarch butterfly, Jeff Foott. 61: Monarch butterflies, Jeff Foott/Bruce Coleman, Inc. 62: Grunion (close-up), Jeff Foott. Grunion, Jeff Foott. 63: Pacific salmon, Jeff Foott. Wolf and salmon, Rollie Ostermick. 64-65: Blue whales, Russ Kinne. 66-67: Wildebeest migration, Sven O. Lindblad from National Audubon Society Collection/P.R. 68: Woodchuck, Leonard Lee Rue III/Bruce Coleman, Inc. Little brown myotis bats, Lynn M. Stone. Prairie rattlesnakes, Tom McHugh from National Audubon Society Collection/P.R. 69: Chipmunk, Breck P. Kent/Animals Animals. 70: Arctic tern chicks, Alvin E. Staffan. 71: Arctic tern, Gordon Langsbury/Bruce Coleman, Inc. 72-73: Canada geese, Ernest D. Feller/Tom Stack & Associates. 73: Canada geese, Mark Newman/Animals Animals. Wandering albatross, Francisco Erize/Bruce Coleman, Inc. Mallard ducks, Bill Wilson from National Audubon Society Collection/P.R. 74-75: Migration map, Bob Hynes.

The Builders
Page 77: Termite mounds, Vincent Serventy/Bruce Coleman, Inc. 78: Cactus wren nest, Jen and Des Bartlett/Bruce Coleman, Inc. Winter wren leaving nest, Eric Soothill/Bruce Coleman, Inc. Blue-gray gnatcatchers and nest, Ron Austing/Bruce Coleman, Inc. Red-capped robin with young in nest, Hans and Judy Beste. Masked weaver nests, Keith Gunnar/Bruce Coleman, Inc. Black-chinned hummingbirds in nest, John Hoffman/Bruce Coleman, Inc. Chestnut-sided warbler on nest, Hal H. Harrison. American robin nest, Jen and Des Bartlett/Bruce Coleman, Inc. 79: Bald eagle and nest, Charles G. Summers, Jr. 80: Burrowing owls, Rod Allin/Bruce Coleman, Inc. 81: Kingfisher, Ron Austing. Pileated woodpecker, James H. Carmichael, Jr. 82: Wilson's phalarope, Alice B. Kessler. 82-83: Flamingos, Roger Tory Peterson. 84: Gannets, Fred Bruemmer. 85: Black storks, M. Philip Kahl/Verda International Photos. 86: Eastern chipmunk, John Shaw. Raccoons, Zig Leszczynski/Animals Animals. 87: Cougar kitten, Leonard Lee Rue III/Animals Animals. 88: Beaver lodge and dam, Greg Beaumont. 89: Lodge illustration, Bob Hynes. Beaver and kits, Jen and Des Bartlett/Bruce Coleman, Inc. 90: Orb weaver and web, Perry Shankle, Jr. 91: Luna moth caterpillar, E.R. Degginger. Caterpillar in leaf, Harry Rogers. 92: Male sockeye salmon, Jeff Foott. 93: Female sockeye salmon, Jeff Foott. Chambered nautilus, Douglas Faulkner/Sally Faulkner Collection. Chambered nautilus illustration, Bob Hynes. Three-spined sticklebacks, Dwight R. Kuhn. 94: White storks, M. Philip Kahl/Photo Researchers. 95: Barn swallows, Laura Riley/Bruce Coleman, Inc. Wasp nest, John Markham/Bruce Coleman, Inc. 96: White-footed mouse, L. West. Young cuckoo and reed warbler, John Markham/Bruce Coleman, Inc. 97: Land hermit crab, James H. Carmichael/Bruce Coleman, Inc.

Finding a Workable Lifestyle
Pages 98-99: Walruses, Leonard Lee Rue III/Animals Animals.

Cooperators and Loners
Page 101: Wolves chasing moose, Rolf O. Peterson. 102: Leafcutter ants, Raymond A. Mendez/Animals Animals. Army ants, Carl W. Rettenmeyer. 103: Honeybees, Ross E. Hutchins. 104-105: Army ants, Carl W. Rettenmeyer. 105: Prairie dog town illustration, Bob Hynes. 106-107: Black-striped grunts, Tom Stack/Tom Stack & Associates. Giraffes, Tom Nebbia. 108-109: Elephants, Wolfgang Bayer. 110: Musk oxen, Fred Bruemmer. 111: Canada geese, Grant Heilman. 112-113: Pelicans,

rwin and Peggy Bauer. 113: elephant dying, Horst Munzig/ Woodfin Camp, Inc. 114: Impala and red-billed oxpecker series, Carl H. Maslowski. 115: Nassau grouper, Douglas Faulkner/Sally Faulkner Collection. 116: Tarantula and anole, Tom McHugh from National Audubon Society Collection/P.R. 116-117: Leopard chasing bush pig, Warren Garst/ Tom Stack & Associates. 117: eastern river snake and green frog, Breck P. Kent/Animals Animals. 118: Bear and cubs, Stouffer Productions/Animals Animals. Anole, James H. Robinson, 119: Olive ridley sea turtle, C. Allan Morgan.

ending and Receiving Messages
Pages 120-121: Red foxes, Larry Lumiller.

The Communicators
Page 123: Mandrill, George H. Harrison. 124: Prairie dogs, David C. Fritts/Animals Animals. Elephant and calf, Des Bartlett/Bruce Coleman, Inc. 125: Rhesus monkeys, Belinda Wright. 126: Dragonfish, William H. Amos. Hatchetfish, William H. Amos. 127: Firefly, L.R. Degginger. Glow-worms, Kjell Sandved, Smithsonian Institution. 128: Hippo, F.S. Mitchell/Tom Stack & Associates. 129: Motmot, M. Philip Kahl/Photo Researchers. Crocodile, George Harrison/Bruce Coleman, Inc. Prairie dog, John M. Burnley from National Audubon Society Collection/P.R. 130-131: White-sided dolphins, Tom McHugh from National Audubon Society Collection/P.R. 132-133: Luna moth, Ray R. Kriner/Grant Heilman Photography. 133: Red wolf, C.C. Lockwood. Sea lion and pup, Francisco Erize. 134: Turkeys, Jerry Shankle, Jr. 134-135: Anole, James H. Robinson. 135: Frog, Hans Pfletschinger/Peter Arnold, Inc. Frigate bird, Leonard Lee Rue III/Bruce Coleman, Inc. 136-137: Japanese red-crested cranes, Tsuneo Hayashida/Orion Press. Yellow-billed storks, M. Philip Kahl/Verda International Photos. 138: Wild dog chasing hyena, Candice Bayer. Rams, John S. Crawford. 139: Giraffes, Simon Trevor/Bruce Coleman, Inc.

Passing on the Secrets of Survival
Pages 140-141: Cheetah family, Wolfgang Bayer.

The Courtship Strategists
Page 143: Sockeye salmon, Rollie Ostermick. 144: Yellow-headed blackbird, Larry R. Ditto. 144-145: Ruffed grouse, Tom and Pat Leeson. 146: Elks, Harry Engels. 147: Golden pheasants, Bruce Coleman/ Bruce Coleman, Inc. Elephant seals, Michael Tennesen. 148: Peacock, Keith Gunnar/Bruce Coleman, Inc. 149: Tree frog, Eric Lindgren/Retna Ltd. Great grey bower bird, Hans and Judy Beste/ Animals Animals. 150: Green orb spiders, Jane Burton/Bruce Coleman, Inc. Lions, Jack Couffer/ Bruce Coleman, Inc. 151: Western grebes, Jen and Des Bartlett/ Bruce Coleman, Inc. Masked boobies, Tui A. De Roy/Bruce Coleman, Inc. 152: Crowned cranes, R.M. Bloomfield/Retna Ltd. Cattle egrets, William J. Weber. 153: Siamese fighting fish, Oxford Scientific Films/Animals Animals. 154: European common frogs, Jane Burton/Bruce Coleman, Inc. 155: Brook trout laying eggs, Stouffer Productions/Animals Animals. Brook trout fertilizing eggs, Stouffer Productions/Animals Animals. 156: Earthworms, Oxford Scientific Films/Animals Animals. Grizzly bears, Stouffer Productions/Animals Animals. Icarus blue butterflies, Peter Ward/Bruce Coleman, Inc. 157: Giant tortoises, Tui A. De Roy/Bruce Coleman, Inc.

The Parents
Page 159: Rockhopper penguins, Grant M. Haist. 160: Whitetail deer giving birth, William J. Weber. 161: Whitetail deer and fawn, William J. Weber. 162: Gadwall duck, R.L. Kothenbeutel. 163: Dormouse, Hans Reinhard/Bruce Coleman, Inc. Hippo and young, M. Erner/Tom Stack & Associates. Purple heron, Udo Hirsch. 164: Scorpion and young, Zig Leszczynski/Animals Animals. 165: Gray kangaroo and joey, Hans and Judy Beste/Tom Stack & Associates. Crayfish, Breck P. Kent/Animals Animals. Baboon and young, Timothy Ransom/Woodfin Camp, Inc. 166-167: Emperor penguins

and young, Arthur Holtzman/ Animals Animals. 167: Canada goose and young, Charles Palek/ Animals Animals. 168: Blue tits and nestlings, U. Berggren/Retna Ltd. Screech owl and owlets, Torrey Jackson. 168-169: Pelican and chick, Harry Engels. 169: Tern and chick, Joe McDonald. 170: Whitetail deer and fawn, William J. Weber. African wild dog and pups, Warren and Genny Garst/ Tom Stack & Associates. 171: Japanese macaque and young, Kojo Tanaka/Animals Animals. 172: Polar bear and cub, Jack W. Lentfer. 173: Giraffe and calf, Donna Grosvenor/Woodfin Camp, Inc. 174: Bison and calf, Gary Randall/ Tom Stack & Associates. 175: Red fox and kit, Wolfgang Bayer/ Bruce Coleman, Inc. Lioness and cubs, Simon Trevor/Bruce Coleman, Inc. Chimpanzee and young, Helmut Albrecht/Bruce Coleman, Inc. 176: Red fox kits, Stephen J. Krasemann/DRK Photo. Lion cubs, M. Philip Kahl/Photo Researchers. 177: Alaskan brown bears, Thase Daniel. 178: Alaskan brown bear and cubs, Tom Bledsoe. 179: Llama and calf, Breck P. Kent/Animals Animals. Lioness and cubs, Baron Hugo van Lawick.

Pushing the Limits
Pages 180-181: Goosefish, Fred Bavendam.

The Innovators
Page 183: Fishing spider, John H. Gerard. 184: Trapdoor spider and beetle, A. Cosmos Blank from National Audubon Society Collection/P.R. 184-185: Jackson's chameleon, Alan Blank/Bruce Coleman, Inc. Bat catching frog, M.D. Tuttle. 185: Loggerhead shrike with cicada, Robert H. Wright from National Audubon Society Collection/P.R. 186: Electric ray, Howard Hall/Tom Stack & Associates. Geometer moth larva James H. Robinson. Echidna, Tom McHugh from National Audubon Society Collection/P.R. 187: Puffer fish, Jane Burton/ Bruce Coleman, Inc. 188: Flying squirrel, Stouffer Productions/ Animals Animals. 189: Walking catfish, Tom Myers. Four-winged

flying fish, Jane Burton/Bruce Coleman, Ltd. 190: Caddis fly larvae, E.S. Ross. Harvest mouse, Ian Beames/Retna Ltd. Water spider, Oxford Scientific Films/Animals Animals. 191: Sociable weaver nest, Clem Haagner/Bruce Coleman, Inc. 192-193: Orangutan, Tom McHugh from National Audubon Society Collection/P.R. 193: Giant panda, George Holton from National Audubon Society Collection/P.R. Clown fish and sea anemone, C.B. Frith/Bruce Coleman, Inc. Ants and aphids, Stephen J. Krasemann/DRK Photo. 194: Hatchetfish, Aldo Margiocco. 195: Proboscis monkey, Miriam Austerman/Animals Animals. Bee dance illustration, Bob Hynes. 196: Ostrich, Fran Allan/Animals Animals. 196-197: Praying mantises, P.H. Ward/Natural Science Photos. 197: Fiddler crab, John Shaw. 198: Nile crocodile with young, Wolfgang Bayer. Mallee fowls, Harold J. Pollock. 199: Sea horse, Jen and Des Bartlett/Bruce Coleman, Inc. 200-201: Giant tortoise, Tui De Roy Moore.

Library of Congress Cataloging in Publication Data

Robinson, David F., 1932-
Living wild.

Includes index.
1. Animals, Habits and behavior of. I. Title.
QL751.R68
591.5
ISBN 0-912186-37-2
     80-80702

**National
Wildlife
Federation**

1412 16th St., N.W.
Washington, D.C. 20036

Thomas L. Kimball
Executive Vice President

J. A. Brownridge
Administrative Vice President

James D. Davis
Director, Book Development

Staff for this Book

Alma Deane MacConomy
Editor

Leah Bendavid-Val
Art Editor

Catherine D. Hughes
Research Editor

Barbara Peters
Writer-Editor

Howard Robinson
Writer-Editor

Robyn Gregg
Editorial Assistant

Dr. Raymond E. Johnson
Wildlife Consultant

David M. Seager
Art Director

Bob Hynes
Illustrator

Priscilla Sharpless
Production and Printing

Patti Matsos
Production Artist

Cathy Pelletier
Permissions Editor

Acknowledgments

The expertise of many wildlife professionals has been an invaluable bulwark to the editors of this book. Among hundreds of sources tapped in its planning and production, the following specialists deserve special thanks for generous help from their wide-ranging fields of observation and study:

Dr. William Shaw, National Oceanic and Atmospheric Administration; Dr. Howard Topoff, Museum of Natural History, N.Y.C.; Drs. R. E. Crabill, Charles Potter, George Watson, Bruce Collett, John Eisenburg, David Nickle, Michael Bogan, and Carl Kranz, Smithsonian Institution; Dr. Merlin Tuttle, Museum of Milwaukee; Dr. Forrest Wood, Naval Ocean System Center, San Diego; Dr. Carl Rettenmeyer, University of Connecticut; Dr. Dewey Caron,

University of Maryland; Mr. Craig Phillips, National Aquarium; Dr. Mark Fuller, Patuxent Wildlife Research Center; Dr. Gil Voss, University of Miami Marine Lab; Dr. Austin McGill and John D. Kaylor, National Marine Fisheries Service.

Thanks go also to the photographers whose work appears in this volume for their helpful observations and accounts of their subjects.

In our own NWF community, thanks are due to Craig Tufts, naturalist, William S. Clark and his Raptor Information Center staff, the library staff, and the editors of *Ranger Rick Nature Magazine, National Wildlife,* and *International Wildlife* magazines. And, as always in our northern Virginia location, we are indebted to the Fairfax County Libraries for their splendid services.